Accelerating and Vortex Laser Beams

Accelerating and Vortex Laser Beams

V. V. Kotlyar

Image Processing Systems Institute of the Russian Academy of Sciences
Branch of the FSRC "Crystallography and Photonics"
Molodogvardeyskaya st. 151, Samara, 443001, Russian Federation

A. A. Kovalev

Samara National Research University
Moskovskoe shosse, 34, Samara, 443086, Russian Federation

CRC Press
Taylor & Francis Group
Boca Raton London New York

CISP

CRC Press is an imprint of the
Taylor & Francis Group, an **informa** business

Translated from Russian by V.E. Riecansky

CRC Press
Taylor & Francis Group
6000 Broken Sound Parkway NW, Suite 300
Boca Raton, FL 33487-2742

First issued in paperback 2021

© 2019 by CISP
CRC Press is an imprint of Taylor & Francis Group, an Informa business

No claim to original U.S. Government works

ISBN 13: 978-1-03-223904-0 (pbk)
ISBN 13: 978-0-367-32236-6 (hbk)

Contents

Contents

Introduction

The exact solutions of Maxwell's equations or other equations of optics – the Helmholtz equation, the paraxial propagation equation of the Schrödinger type – have always attracted the attention of researchers. These solutions describe electromagnetic coherent light fields and laser beams, which are widely used in practice and whose properties can be described in detail analytically. Such well-known light fields that have an exact analytical description include: plane wave, spherical wave, Gaussian, and Bessel beams [1]. Some time ago, new light beams that had an accurate analytical description were discovered: Hermite–Gauss and Laguerre–Gauss mode beams [2], Hermite–Laguerre–Gauss beams [3], Mathieu [4] and Ince–Gaussian [5] beams, [6], accelerated Airy beams [7], Pearcey self-focussing beams [8], and others. Light fields can be divided into two classes having orbital angular momentum (OAM) [9] and not having it. Beams with OAM are called vortex or singular. The vortex laser beams [10] have such distinctive features: a helical or spiral phase, wave-front dislocations, isolated points of zero intensity. The vortex laser beams have recently become widespread, they are used in probing the atmosphere in the presence of turbulence [11], in wireless communication systems [12], to condense information transmission channels through fibers [13], in astronomy [14], quantum computer science [15] and micromanipulations [16].

The authors have previously written about vortex laser beams [10], but over the past several years, new vortex beams have appeared: asymmetric Bessel and Laguerre–Gauss beams, Hermite–Gauss vortex beams, Lommel modes, half Pearcey beams, vector vortex Hankel beams and others. This book is devoted to their consideration.

The authors are grateful to Dr. A.P. Porfirev, the research associate of the Samara University, for conducting experimental studies of vortex laser beams and Dr. A.G. Nalimov, researcher at the Image

Processing Systems Institute, the Russian Academy of Sciences, for simulation of the focussing of light using gradient elements of micro-optics.

This book was supported by a grant from the Russian Foundation for Basic Research 18-29-20003.

Accelerating laser beams

1.1. Airy beams with a non-parabolic trajectory

1.1.1. Airy beams with a parabolic trajectory

In 1974 Kalnins and Miller [17] found a solution to the paraxial propagation equation in the form of Airy functions. In 1979, Berry and Balazs [18] considered non-broadening wave packets in quantum mechanics. In 1994, Besieris et al. [19] found a solution to the 2D paraxial equation

$$2i\frac{\partial U}{\partial \xi} + \frac{\partial^2 U}{\partial s_x^2} + \frac{\partial^2 U}{\partial s_y^2} = 0 \qquad (1.1)$$

as

$$U(s_x, s_y, \xi) = \mathrm{Ai}\left(s_x - \frac{\xi^2}{4}\right)\mathrm{Ai}\left(s_y - \frac{\xi^2}{4}\right)\exp\left(\frac{is_x\xi}{2} + \frac{is_y\xi}{2} - \frac{i\xi^3}{6}\right), \qquad (1.2)$$

where $\mathrm{Ai}(x)$ is the Airy function, equal for real positive values of the argument

$$\mathrm{Ai}(x) = \pi^{-1}(x/3)^{1/2} K_{1/3}\left(2x^{3/2}/3\right) = \\ \sqrt{x}\left[I_{-1/3}\left(2x^{3/2}/3\right) - I_{1/3}\left(2x^{3/2}/3\right)\right]/3, \qquad (1.3)$$

$I_{1/3}(x)$ and $K_{1/3}(x)$ are modified Bessel functions of the first and second kind, $s_x = x/x_0$, $s_y = y/x_0$, $\xi = z/(kx_0^2)$ are the dimensionless transverse and longitudinal Cartesian coordinates, $k = 2\pi/\lambda$ is the wave number, x_0 is an arbitrary transverse size. Airy beams (AB) in [17–19] have infinite energies, since the Airy function slowly decreases as $x \to \infty$:

$$\mathrm{Ai}(-x) \approx \pi^{-1/2} x^{-1/4} \sin\left[(2/3)x^{3/2} + (\pi/4)\right]. \tag{1.4}$$

In 2007, Siviloglou and Christodoulides [7] examined AB in finite-energy optics. They showed that by solving equation (1.1) in the 1D variant

$$2i\frac{\partial U}{\partial \xi} + \frac{\partial^2 U}{\partial s^2} = 0 \tag{1.5}$$

has the appearance

$$U(s,\xi) = \mathrm{Ai}\left(s - \frac{\xi^2}{4} + ia\xi\right)\exp\left(\frac{is\xi}{2} + \frac{ia^2\xi}{2} - \frac{i\xi^3}{12} - \frac{a\xi^2}{2} + as\right), \tag{1.6}$$

which in the initial plane with $\xi = 0$ gives the Airy function with exponential apodization

$$U(s,\xi=0) = \mathrm{Ai}(s)\exp(as), \quad a > 0, \tag{1.7}$$

where a is a constant. It was also shown in [7] that an 1D AB beam with a finite energy can be formed by illuminating a phase mask with a cubic dependence on the transverse coordinate by a Gaussian beam and performing a Fourier transform with a spherical lens. This follows from the fact that the Fourier transform of the initial field (1.7) has the form:

$$F(t) = \exp(-at^2)\exp\left(\frac{it^3}{3} - ia^2t + \frac{a^3}{3}\right). \tag{1.8}$$

The main feature of AB is the curvature of the trajectory of the main maximum (main lobe). It propagates along a parabolic trajectory. Therefore, ABs are called accelerating or ballistic (since a freely falling body moves along a parabola with acceleration g). However, in [20] it is shown that in the ABs with a finite energy (1.6), (1.7) 'beam's centre of gravity' is not displaced during propagation and the accelerating effect is manifested only for small values of $a \ll 1$. In experiments [21], using a liquid-crystal light modulator, a parabolic propagation trajectory of 1D AB was obtained, which was well described by the formula ($a = 0.1$):

$$x = \left(\frac{\lambda z}{4\pi x_0^{3/2}}\right)^2.$$

(1.9)

where (x, z) is the transverse and longitudinal Cartesian coordinates. In [22], a generalization of AB – Airy–Gauss (AG) beams was obtained and their transformation was calculated in an ABCD optical system. If ABs with infinite energy in the initial plane in [17, 18] have the form $E_0(x, z = 0) = \text{Ai}(x)$, and the finite-energy beams from [7] have the form (7) in the initial plane: $E_0(x, z = 0) = \text{Ai}(x)\exp(ax)$, then the AG-beams from [22] in the initial plane have the following form:

$$E_2(x, z = 0) = \text{Ai}\left(\frac{x+\delta}{\beta}\right)\exp\left[(a+ib)x - \frac{x^2}{w^2} + \frac{ikx^2}{2R}\right],$$

(1.10)

where δ, β, a, b, w, R are parameters. In [23] it was shown that the addition of a linear tilt to the AB [17] allows one to control the parameters of a parabolic trajectory. An expression for the Fresnel transformation from the AB in the initial plane has the form

$$E_3(s, z = 0) = \text{Ai}(s)\exp(as + ivs), \quad s = x/x_0,$$

(1.11)

and it is shown that the trajectory of such a beam, in contrast to (1.9), has the form

$$x = \frac{vz}{kx_0} + \frac{z^2}{4k^2 x_0^3}.$$

(1.12)

In [24], the Poynting vector and the orbital angular momentum for the AB were calculated numerically. In [25] Bandres considered accelerated parabolic beams. In the initial plane, their complex amplitude has the form:

$$E_4(u,v,z=0) = \theta_n(\eta)\theta_n(i\xi)\exp\left[iw(\eta^2-\xi^2)/2+i\xi^3/3\right], \qquad (1.13)$$

where ξ and η are the parabolic coordinates:

$$\begin{cases} u-\lambda-w^2 = \left(\eta^2-\xi^2\right)/2, \\ v = \eta\xi, \end{cases} \qquad (1.14)$$

and the functions $\theta_n(x)$ satisfy the equation of the anharmonic oscillator:

$$\left(-\frac{1}{2}\frac{\partial^2}{\partial\eta^2}+\frac{\eta^4}{4}\right)\theta_n(\eta) = E_n\theta_n(\eta). \qquad (1.15)$$

For the function $\theta_n(x)$, the analytical solution in the form of special functions is not known, only the asymptotics is known:

$$\theta_n(i\xi) \approx \sin(6+\pi/6)/\xi, \quad \xi\to\infty. \qquad (1.16)$$

In [26] experiments were carried out using a 35 fs pulse at an average wavelength $\lambda = 0.8$ µm with an energy of 10 mJ in air with multiphoton ionization to produce a curved plasma channel with a radius of about $\sigma = 132$ µm and a length of 69 cm, which is satisfactorily described by the formula:

$$x = 0.037\frac{\lambda^2 z^2}{\sigma^3}, \qquad (1.17)$$

where σ is the radius of the Gaussian beam. In [27] Bandres considered another type of AB – accelerated beams. He found a solution to the equation

$$\left(i\frac{\partial}{\partial\xi}+\frac{\partial^2}{\partial u^2}+\frac{\partial^2}{\partial v^2}\right)E(u,v,\xi) = 0 \qquad (1.18)$$

as

$$E_5(u,v,\xi) = \text{Ai}(u+\omega^2)\exp\left[i\xi(u-\lambda-\xi^2)+i\xi^3/3\right]\exp(i\omega v), \qquad (1.19)$$

where λ, ω are constants. Although this beam is two-dimensional, but along one coordinate its trajectory is parabolic as in 1D AB. In [28] it is shown numerically in the nonparaxial case that a 2D field with a cubic phase bounded by a circular diaphragm with a radius R

$$E_x(x,y,z=0) = \mathrm{circl}\left(\frac{\sqrt{x^2+y^2}}{R}\right)\exp\left[i\beta(x^3+y^3)\right] \qquad (1.20)$$

also has a parabolic trajectory before and after the Fourier plane. In [29], using 3-wave mixing in a nonlinear photonic crystal and using temperature control, the authors obtained two modes (second harmonic generation and differential frequency generation), in which the initial cubic wave front generated AB with parabolas directed in different directions. Similar to [28] in [30] 1D AB bounded along the transverse coordinate and a linear combination of AB are numerically investigated, which, when propagated, forms two parabolas, pointing in different directions, similarly to [29]. In [31], the effect of the mutual displacement of the centre of the Gaussian beam and the centre of the cubic phase mask on the formation of modernized ABs in the Fourier plane is investigated. An explicit expression for the Fourier transform of the initial function is obtained.

$$E_6(t) = \exp\left[-a(t-t_0)^2\right]\exp\left\{i\left[\frac{(t-t_1)^3}{3} - a^2(t-t_1) - i\frac{a^3}{3}\right]\right\}. \qquad (1.21)$$

It is shown that the choice of the constants t_0 and t_1 can control the shape of the parabolic trajectory of AB. In [32], another type of finite-energy 1D AB bounded on one side along the transverse coordinate is studied. In the initial plane, these ABs have the following complex amplitude:

$$E_7(s,\xi=0) = \mathrm{Ai}_0(s) \pm i\,\mathrm{Bi}_1(s), \qquad (1.22)$$

where $\mathrm{Ai}(x)$ and $\mathrm{Bi}(x)$ are two linearly independent solutions of the Airy equation

$$\left(\frac{\partial^2}{\partial x^2} - x\right)E(x) = 0, \qquad (1.23)$$

but limited on the one hand in a special way:

$$\mathrm{Ai}_0(s) = \begin{cases} \mathrm{Ai}(s)\exp(as), s < s_0 = -2,3381, \\ 0, \quad s \geq s_0, \end{cases} \qquad (1.24)$$

$$\mathrm{Bi}_1(s) = \begin{cases} \mathrm{Bi}(s)\exp(as), s < s_1 = -1,1737, \\ 0, \quad s \geq s_1. \end{cases} \qquad (1.25)$$

The distribution of such modernized ABs is investigated numerically in [32]. In [33], a radially symmetric AB with finite energy was proposed, which 'self-focused' at a certain distance:

$$E_8(r) = \mathrm{Ai}(r - r_0)\exp\left[a(r - r_0)\right], \qquad (1.26)$$

where r and r_0 are variable and constant transverse radial coordinates. The focal length depends on the value of r_0. The propagation of a beam is modelled numerically. It was theoretically shown in [34] that using a thin wedge-shaped crystal, whose thickness varies linearly along the transverse coordinate, and when pumped with a Gaussian beam near the exit surface of the crystal, cut AB with acceleration occurs (with a parabolic motion path). B [35] considers the AB called the 'light bullet' numerically and experimentally. In the initial plane at any time, its complex amplitude has the form:

$$E_9(x,y,z,t) = \mathrm{Ai}(x/x_0)\mathrm{Ai}(y/y_0)\mathrm{Ai}\left(\tau/\tau_0 - \frac{\beta z^2}{4\tau_0^4}\right), \qquad (1.27)$$

where $\tau = t - z/c$, x_0, y_0, τ_0 and β are constant.

1.1.2. Beams accelerating along a non-parabolic trajectory

In [36], using the stationary phase method from the Fresnel integral, an expression for the complex field amplitude was obtained, which focuses on a curve (the stationary points of which lie on this curve) $x = f(z)$:

$$E_{10}(x,z) = \sum_{n=1}^{\infty} C_n \mathrm{Ai}\left[\frac{k^{2/3}\left(\varphi''(x_{0n}) + 1/z\right)^2}{2^{2/3}\left(\varphi'''(x_{0n})\right)^{4/3}}\right], \qquad (1.28)$$

where φ, φ'', φ''' are the phase functions of the field in the initial plane $z = 0$ and its second and third derivatives, x_{0n} are the stationary points of the exponent in the Fresnel integral, equal to $k\varphi + k(x - x_0)^2/(2z)$, where x_0 is the coordinate in the initial plane. These stationary points satisfy the condition $x = x_{0n} + z\varphi'(x_{0n})$. With the help of formula (1.28) and the spatial light modulator, experimentally obtained ABs whose trajectories are described by the polynomial $x = z^m$, where $m = 1.5; 2; 3; 4; 5$. In [37], a light field is considered that propagates in a planar gradient inhomogeneous medium with a linear distribution of the refractive index and satisfies the paraxial equation in dimensionless units (similar to equation (1.5)):

$$\left(i\frac{\partial}{\partial z} + \frac{1}{2}\frac{\partial^2}{\partial x^2} - \frac{\alpha(z)x}{2} \right) E(x,z) = 0, \qquad (1.29)$$

where $\alpha(z)$ is a function of the longitudinal variable z. The solution of equation (1.29) is obtained in the form of integrals for any $\alpha(z)$ and with an initial field of the form $E_0(x, z = 0) = \text{Ai}(\gamma^{1/3}x)$, where γ is the scaling factor. In [38], an equation for the phase $\varphi(x)$ of the light field was obtained, which directs the rays to the 1D caustic curve $c(z)$:

$$\frac{d\varphi(x)}{dx} = \frac{k[c(z) - x]}{\sqrt{[c(z) - x]^2 + z^2}}, \qquad (1.30)$$

where k is the wave number of light. From this equation it follows that with $c(z) = az^2$ phase is obtained in degree 3/2, and the amplitude AB in the initial plane has the form:

$$E_{11}(x, z = 0) = \exp\left(-i4\sqrt{a}kx^{3/2}/3\right), \qquad (1.31)$$

where a is a constant. If the caustic equation is $c(z) = az^4$, then the initial field will be as follows:

$$E_{12}(x, z = 0) = \exp\left(-i16(3a)^{1/3}kx^{7/4}/21\right). \qquad (1.32)$$

Similarly, an explicit phase for the caustic $c(z) = az^m$ was found in [38], where m is an integer.

In [39], the acceleration of AB using gradient optics was studied theoretically and experimentally. For a paraxial equation with a weak linear refractive index:

$$\left(i\frac{\partial}{\partial z} + \frac{1}{2k}\frac{\partial^2}{\partial x^2} - \frac{k\,\delta\,x}{2n} \right) E(x,z) = 0, \tag{1.33}$$

where $\delta \cdot x \ll n$, n is the refractive index of the medium, an explicit solution is obtained in the form:

$$E_{13}(x,z) = \mathrm{Ai}\left[\frac{1}{x_0}\left(x - \frac{z^2}{4k^2 x_0^3} - \frac{\delta z^2}{2n} \right) + i\frac{az}{kx_0} \right] \exp \times$$

$$\times\left[ax - \frac{az^2}{2}\left(\frac{1}{k^2 x_0^3} + \frac{\delta}{n} \right) \right] \times$$

$$\times \exp\left[-\frac{iz^3}{12}\left(\frac{1}{k^3 x_0^6} + \frac{2k\delta^2}{n^2} + \frac{3\delta}{nkx_0^3} \right) + \frac{ia^2 z}{2k} \right] \exp \times \tag{1.34}$$

$$\times\left[ixz\left(\frac{1}{2kx_0^3} + \frac{k\delta}{n} \right) \right],$$

satisfying the boundary condition $E(x,\ z = 0) = \mathrm{Ai}(x/x_0)\exp(ax)$, $a > 0$. In [40], plasmon ABs were experimentally obtained using a matrix of subwave holes in a silver film, periodic along the z axis and non-periodic (approximating a phase of degree 3/2) along the x axis. The plasmon wave incident on the holes propagates along the x axis, and AB propagates along the z axis. In [41], AB was formed using a liquid crystal modulator whose transmission is described by a phase function of degree 3/2: $\varphi(x) = ax^{3/2}$. In [42] considered 2D triple ABs in the initial plane in the form:

$$E_{14}(x,y,z=0) = \mathrm{Ai}(by+c)\,\mathrm{Ai}\left(b\frac{x\sqrt{3}-y}{2}+c \right)\mathrm{Ai}\left(b\frac{-x\sqrt{3}-y}{2}+c \right). \tag{1.35}$$

This beam has finite energy. The Fourier transform of this beam was found analytically in [42], and the Fresnel transform was calculated numerically. In [43], a reflecting diffraction grating with a relief was used as a mirror in an external Nd:YAG resonator:

$$T(x,y) = \frac{h_0}{2}\left[\text{sign}\left\{ \cos\left(\frac{2\pi x}{\lambda} + \frac{x^3}{a} + \frac{y^3}{a} \right) \right\} + 1 \right],\tag{1.36}$$

where h_0 is the height of the lattice teeth, a is the scaling factor.

In this case, the laser generates a 2D AB. In [44], in the integral form for the Bessel function, it was proposed to integrate only over half a circle from zero to π:

$$J_v^+(x,z) = \int_0^\pi \exp\left[ivt + ik(x\cos t + z\sin t) \right]dt.\tag{1.37}$$

Formula (1.37) describes a 1D accelerating beam propagating along a circular path with a rotation of almost 90 degrees. In [45], a 2D Bessel beam was considered, but it was interpreted as an 1D beam propagating along the z axis:

$$E_{15}(x,z) = J_v\left(k\sqrt{x^2 + z^2} \right)\exp\left(iv\arctan\frac{z}{x} \right),\tag{1.38}$$

where $J_v(x)$ is the Bessel function. And at $z = 0$ is selected only on the semiaxis:

$$E_{15}^+(x,z=0) = J_\beta(kx+\beta)\exp(-ax)\theta(x+\beta/k),\tag{1.39}$$

where $\theta(x)$ is the Heaviside function. The accelerating beam (1.39), like the beam (1.37), rotates when it propagates around the circle by 90 degrees. In [46] experimentally using a modulator and a laser pulse with a duration of 10 fs a light beam with a caustic curve in the form of a circle with a radius of 35 μm was obtained. At the same time a phase function was formed on a liquid crystal modulator, calculated using the equation of geometric optics, similar to equation (1.30):

$$\frac{d\varphi(x)}{dx} = \frac{kc'}{\sqrt{1+(c')^2}}, \quad c' = \frac{dc}{dz}.\tag{1.40}$$

In (1.40), the function $c(x)$ defines the desired caustic curve, for example, an arc of a circle.

In [47], a circular AB with an optical vortex with an initial amplitude was considered:

$$E_{16}(r,\varphi,z=0)=\text{Ai}\left(\frac{r-r_0}{w}\right)\exp\left[a\frac{r-r_0}{w}\right]\left[r\exp(i\varphi)-r_0\exp(i\varphi_0)\right]^n, \quad (1.41)$$

where n is the topological charge of the vortex. This beam forms a light ring in the focus. In [48], nonparaxial corrections to paraxial AB were found using the virtual source method. In [49], proceeding from the 2D Mathieu beam, taking its one-dimensional section, laser beams accelerated in elliptical orbits were considered. The Mathieu beam is selected in the form:

$$E_{17}(u,v,z=0)=A\text{ce}_m(v)\text{Mc}_m(u)+iB\text{se}_m(v)\text{Ms}_m(u), \quad (1.42)$$

where A, B are constants, (u, v) are the elliptical coordinates $x = f$ ch(u) cos(v), $y = f$ sh(u)sin(v), ce, se are the angular and Mc, Ms are the Mathieu radial functions. For an accelerating beam, we must put $v = 0$ (or $v = \pi/2$).

In the above detailed review of works on accelerating beams and, in particular, ABs, there are no papers on ABs, which are accelerated along hyperbolic paths. To fill this problem, we consider further Airy beams of the second kind (AB-2) or hyperbolic AB (HAB).

1.1.3. Airy laser beams of the second kind

In [50], another type of accelerated Airy beams was proposed. Usually, 1D AB is formed using an initial field (1.8) – a cubic phase mask, then a Fourier spectrum of the field (1.8) is formed using a spherical lens, which is described by the function (1.7). And behind the Fourier plane, AB is formed with a complex amplitude (1.6). AB-2 occur in the Fresnel zone of the phase mask (1.8). To show this, consider the complex amplitude of the Gaussian beam immediately behind the cubic phase mask:

$$E(x,0)=\exp\left[-\frac{x^2}{w^2}+i\alpha\left(\frac{x}{x_0}\right)^3+i\beta\left(\frac{x}{x_0}\right)\right], \quad (1.43)$$

where w is the radius of the waist of the Gaussian beam, α and β are the dimensionless parameters of the phase mask. Then, at a distance z from the initial plane in the paraxial approximation, the amplitude of the light field will be described by the Fresnel transformation:

$$E(x,z) = \sqrt{\frac{-ik}{2\pi z}} \exp(ikz) \int\limits_{-\infty}^{+\infty} \exp\left[\frac{ik}{2z}(x-t)^2\right]$$

$$\exp\left[-\frac{t^2}{w^2} + i\alpha\left(\frac{t}{x_0}\right)^3 + i\beta\left(\frac{t}{x_0}\right)\right] dt.$$

(1.44)

Complementing the exponent to the full cube and using the well-known integral representation for the Airy function

$$\text{Ai}(x) = \frac{1}{2\pi} \int\limits_{-\infty}^{\infty} \exp\left(\frac{it^3}{3} + ixt\right) dt,$$

(1.45)

we can calculate the integral in (1.44). Then we get:

$$E(x,z) = \sqrt{\frac{-i2\pi k}{z}} wp \exp\left[\frac{ikx^2}{2z} + sp(qp)^2 + \frac{2}{3}(qp)^6 + ikz\right] \times$$

$$\times \text{Ai}\left[sp + (qp)^4\right],$$

(1.46)

where

$$z_0 = \frac{kw^2}{2}, \quad q^2 = 1 - \frac{iz_0}{z}, \quad s = \frac{w}{x_0}\left(\beta - \frac{kx_0 x}{z}\right), \quad p = \frac{x_0}{w\sqrt[3]{3\alpha}}.$$

(1.47)

In the original notation of (1.43) instead of (1.46) we can write:

$$E(x,z) = \sqrt{\frac{-i2\pi k}{z}} \frac{x_0}{\sqrt[3]{3\alpha}} \times$$

$$\times \exp\left[\frac{1}{3\alpha}\left(\frac{x_0}{w}\right)^2\left(\beta - \frac{kx_0 x}{z}\right) + \frac{2}{27\alpha^2}\left(\frac{x_0}{w}\right)^6\left(1 - 3\frac{z_0^2}{z^2}\right)\right] \times$$

$$\times \exp\left[\frac{ikx^2}{2z} - \frac{iz_0}{z}\frac{1}{3\alpha}\left(\frac{x_0}{w}\right)^2\left(\beta - \frac{kx_0 x}{z}\right)\right] \times \tag{1.48}$$

$$\times \exp\left[-\frac{2i}{27\alpha^2}\left(\frac{x_0}{w}\right)^6\left(3\frac{z_0}{z} - \frac{z_0^3}{z^3}\right) + ikz\right] \times$$

$$\times \text{Ai}\left\{\frac{1}{(3\alpha)^{1/3}}\left[\beta - \frac{kx_0 x}{z} + \frac{1}{3\alpha}\left(\frac{x_0}{w}\right)^4\left(1 - \frac{iz_0}{z}\right)^2\right]\right\}$$

Expression (1.48) describes AB-2 with finite energy. From (1.48) it is clear that AB-2, unlike AB (1.6), has a quadratic phase, and not a linear phase, and therefore it will diverge during propagation. In addition, in (1.48) the argument of the Airy function is complex, as in (1.6), but the dependence on the z coordinate has a different character: in (1.6), the argument value of the Airy function is proportional to z^2, and in (1.48) is inversely proportional to z. AB-2 with infinite energy can be obtained if, instead of a Gaussian beam, we illuminate a cubic phase mask with a plane wave ($w \to \infty$). Then instead of (1.48) we get the expression:

$$E(x,z) = \sqrt{\frac{-i2\pi k}{z}} \frac{x_0}{\sqrt[3]{3\alpha}} \times$$

$$\times \exp\left[\frac{ik}{2z}\left(x^2 + \frac{kx_0^3 x}{3\alpha z} - \frac{\beta x_0^2}{3\alpha} + \frac{k^2 x_0^6}{54\alpha^2 z^2}\right) + ikz\right] \times \tag{1.49}$$

$$\times \text{Ai}\left[\frac{1}{(3\alpha)^{1/3}}\left(\beta - \frac{kx_0 x}{z} - \frac{k^2 x_0^4}{12\alpha z^2}\right)\right].$$

Expression (1.49) describes AB-2 with infinite energy. But the quadratic dependence of the phase is preserved and therefore the beam (1.49) will diverge during propagation. The argument for the

Airy function in (1.49) is a real number. We equate this argument to the value in which the Airy function has local maxima y_m:

$$\frac{1}{\sqrt[3]{3\alpha}}\left(\beta - \frac{kx_0 x}{z} - \frac{k^2 x_0^4}{12\alpha z^2}\right) = y_m. \tag{1.50}$$

The numbers y_m in (1.50) take the following values:

Table 1

m	y_m
0	−1.01879
1	−3.2482
2	−4.8201
3	−6.16331
4	−7.37218
5	−8.48849
6	−9.53545
7	−10.5277
8	−11.4751
9	−12.3848
10	−13.2622

From equation (1.50) you can find the explicit equation of the trajectory of the maximum of AB-2:

$$x = \frac{z}{kx_0}\left(\beta - y_m \sqrt[3]{3\alpha}\right) - \frac{kx_0^3}{12\alpha z}. \tag{1.51}$$

In contrast to the parabolic trajectory (1.9), along which AB propagate, AB-2 along a hyperbolic trajectory.

Find the derivatives on z from the expression (1.51):

$$\frac{dx}{dz} = \frac{1}{kx_0}\left(\beta - y_m \sqrt[3]{3\alpha}\right) + \frac{kx_0^3}{12\alpha z^2}, \tag{1.52}$$

$$\frac{d^2 x}{dz^2} = -\frac{kx_0^3}{6\alpha z^3}. \tag{1.53}$$

The hyperbolic trajectory (1.51) has acceleration in those areas where the derivatives of the first and second order (1.52) and (1.53) have the same sign. It follows that to accelerate the trajectory should be

$$\frac{1}{z^2} < \frac{12\alpha}{k^2 x_0^4}\left(y_m \sqrt[3]{3\alpha} - \beta\right), \tag{1.54}$$

moreover, the condition for acceleration does not depend on the sign of x_0/α. For the existence of distances z satisfying (1.54), the right side of these expressions must be positive. This is possible when $\alpha > 0$ and $\beta < y_m(3\alpha)^{1/3}$, or $\alpha < 0$ and $\beta > y_m(3\alpha)^{1/3}$, i.e.,

$$\mathrm{sign}(\alpha)\beta < y_m \sqrt[3]{3|\alpha|}. \tag{1.55}$$

If the condition (1.55) is satisfied, then acceleration is observed at distances

$$z > z_1 = \frac{k x_0^2}{2\sqrt{3\alpha\left(y_m \sqrt[3]{3\alpha} - \beta\right)}}, \tag{1.56}$$

moreover, unlike Airy beams of the first type, acceleration is not constant, but decreasing in proportion to z^{-3}, starting from

$$\left.\frac{\mathrm{d}^2 x}{\mathrm{d}z^2}\right|_{z=z_1} = \begin{cases} -\dfrac{4\sqrt{3\alpha}}{k^2 x_0^3}\left(y_m \sqrt[3]{3\alpha} - \beta\right)^{3/2}, \alpha > 0, \\[3mm] \dfrac{4\sqrt{-3\alpha}}{k^2 x_0^3}\left(\beta - y_m \sqrt[3]{3\alpha}\right)^{3/2}, \alpha < 0. \end{cases} \tag{1.57}$$

For example, consider the following parameter values: $\lambda = 532$ nm, $x_0 = \lambda$, $\alpha = -1$, $\beta = 10$, $m = 0$, $y_0 = -1.01879$. In this case the condition (1.55) is fulfilled and the trajectory has acceleration at $z > z_1 \approx 330$ nm. The trajectory graph with the specified parameters is shown in Fig. 1.1 *a*, and the intensity distribution of the field (1.49) with the same parameters is shown in Fig. 1.1 *b*. The calculation domain in Fig. 1.1 *b* has dimensions $-10\lambda \le x \le +10\lambda$, $0 \le z \le 4\lambda$.

Figure 1.2 shows the intensity cross section in the planes $z = \lambda/2$ (*a*), λ (*b*), 2λ (*a*), 3λ (*g*), 4λ (*d*).

For comparison, consider the first type Airy beam (1.6) with $a = 0$. Equate in (1.6) the argument of the Airy function to the value in which it has local maxima y_m and find the explicit equation of the trajectory of the maximum AB:

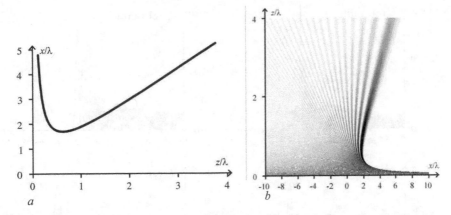

Fig. 1.1. The trajectory of AB-2 with acceleration (a) and the intensity distribution of AB-2 with acceleration in the xz plane (b).

$$x = x_0 y_m + \frac{z^2}{4k^2 x_0^3}. \tag{1.58}$$

From (1.58) it is seen that the AB beam has a constant acceleration equal to $1/(2k^2 x_0^3)$. It is also seen from (1.58) that the AB (1.6) is non-divergent (without diffraction), since $x_1 - x_2 = x_0 (y_m - y_n)$ does not depend on z, and for the AB-2 (1.51) it follows that $x_1 - x_2 = (3\alpha)^{1/3} z (y_m - y_n)/(kx_0)$ and the beam diverges linearly with increasing z.

Figure 1.3 shows the field intensity distribution (1.6) with the following parameters: $\lambda = 532$ nm, $x_0 = \lambda/2$. The calculated domain in Fig. 1.3 has the same dimensions $-10\lambda \le x \le +10\lambda$, $0 \le z \le 4\lambda$. According to (1.58), the acceleration for such a beam is $1/(\pi^2 \lambda)$, whereas, according to (1.57), for the AB-2 beam shown in Fig. 1.1 b, at $z = z_1 \approx 330$ nm (z_1 is derived from (1.56)) is equal to about $19.87/(\pi^2 \lambda)$. This determines the more curved shape of the trajectories in Fig. 1.1 b.

The following results were obtained in the section. A sufficiently detailed review of the scientific work on accelerating the laser beam, including the Airy beams is presented; the review was made in order to prove that the light beams generated by the cubic phase mask in the Fresnel zone have not yet been considered. An explicit form of the complex amplitude, which describes the diffraction of a Fresnel Gaussian beam on the phase mask with the cubic phase dependence hydrochloric transverse coordinates (equation (1.48)). One-dimensional Airy beams of the second kind with finite energy (Eq. (1.48)) and with infinite energy (Eq. (1.49)), which propagate

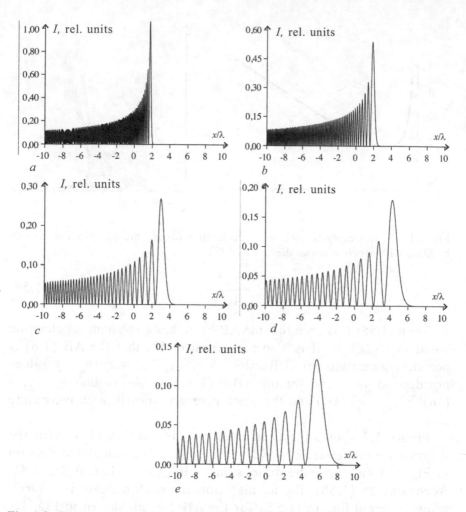

Fig. 1.2. Cross-sections of the intensity of the beam AB-2 in the planes $z = \lambda/2$ (*a*), λ (*b*), 2λ (*c*), 3λ (*d*), 4λ (*e*).

with acceleration on a portion of the hyperbolic trajectory (Eq. (1.57)), are examined. It turned out that second-kind Airy beams have inhomogeneous 'acceleration', which quickly attenuates (proportional to the cube of the distance) and the beam continues to propagate along a straight-line trajectory (Fig. 1.1). It was also shown that second-kind Airy beams in the near zone (several wavelengths from the cubic phase mask) are 'accelerated' by an order of magnitude more (other parameters being equal) than ordinary Airy beams propagating along a parabolic trajectory. The Airy second kind beams with infinite energy propagation diverge (central intensity peak broadened) linearly with distance from the initial plane.

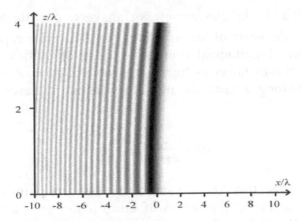

Fig. 1.3. The intensity distribution of AB-1 (with constant acceleration) in the *xz* plane.

1.2. Distance of diffraction-free propagation of the Airy beam

As is known, the Airy beams [7, 18] have infinite energy, and therefore, when propagating along a parabola, they retain their intensity, that is, they propagate without diffraction. But in practice, it is necessary either to limit the Airy beam in the initial plane with the final aperture [51], or to apodize it with an exponential [7] or Gaussian function [52]. At the same time, the Airy beam, in the initial part of its trajectory, retains its remarkable properties (diffractionlessness, propagation along a parabola and self-recovery upon distortion) to a certain distance, and then begins to diverge, and the amplitude at the main maximum begins to decrease to zero. It was shown in [51] that the Airy beam, bounded by the aperture, is much less subject to diffraction than the Airy beam truncated by the Gaussian function, or exponentially. The purpose of this section is to provide a formula for determining the maximum distance on the optical axis (starting from the initial plane) on which the Airy beam with the limited aperture retains the property of invariance.

1.2.1. Getting the distance of invariance from the equation of the trajectory

The Airy beam is described by a complex amplitude [7, 18]:

$$E(s,\xi) = \mathrm{Ai}\left(s - \frac{\xi^2}{4}\right)\exp\left[i\left(-\frac{\xi^3}{12} + \frac{s\xi}{2}\right)\right], \qquad (1.59)$$

where $s = x/x_0$ is the dimensionless transverse coordinate along the x axis, x_0 is the scale of the Airy function $\text{Ai}(x)$, $\xi = z/(k x_0^2)$ is the dimensionless longitudinal coordinate along the optical axis z, $k = 2\pi/\lambda$ is the wavenumber of light with wavelength λ. A beam (1.58) propagates along a parabolic trajectory, which is described by the equation:

$$\Delta x = \frac{z^2}{4k^2 x_0^3},\tag{1.60}$$

where Δx is the transverse displacement of the point of the Airy beam, which was at the origin of coordinates ($x = z = 0$). From (1.60) it follows that of the Airy spot beam which at $z = 0$ had the coordinate $x = -\Delta x$, while another value of z in equation (1.60) will have a coordinate $x = 0$. That is, if we assume that $\Delta x = R$ is the modulus of the coordinate of the edge of the diaphragm (hereinafter, this value will be called the diaphragm radius), which bounds the Airy beam on the negative part of the axis

$$E(x, z = 0) = \begin{cases} \text{Ai}(x/x_0), & x \geq -R, \\ 0, & x < -R, \end{cases}\tag{1.61}$$

then from (1.60) we find the expression for the distance z_1 for which the point with coordinates ($x = -R$, $z = 0$) (diaphragm edge) will shift to a point with the coordinates ($x = 0$, $z = z_1$) (optical axis):

$$z_1 = 2kx_0^{3/2}\sqrt{R}.\tag{1.62}$$

In the derivation of (1.62), we implicitly assumed that the beam emerging from the starting point with the coordinates ($x = -R$, $z = 0$) intersects the parabolic trajectory of the main maximum of the Airy beam at the point with the coordinates ($x = R$, z). Next we prove this hypothesis and show that this distance z_1 is the distance of the Airy beam invariance. Figure 1.4 shows the intensity distribution in the plane Oxz for different values of the radius of the diaphragm. From Fig. 1.4 it is seen that the trajectory of the main maximum forms a caustic, with the beam emerging from the extreme point of the diaphragm ($x = -R$, $z = 0$), extends at an angle β to the optical axis and is tangent to the caustic (Fig. 1.4 b). Let us find the angle β, assuming that $R > 0$. Obviously, the trajectory of the main maximum,

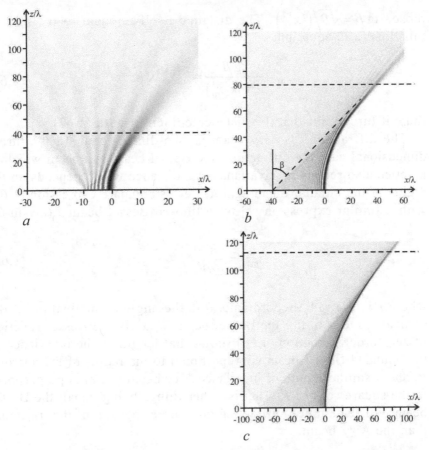

Fig. 1.4. The intensity in the plane *Oxz* for the diaphragm radi $R = 10\lambda$ (*a*), $R = 40\lambda$ (*b*) and $R = 80\lambda$ (*c*). The horizontal dashed line shows the length of the Airy beam invariance calculated by the formula (1.62),

described by the formula (1.59), and the straight line corresponding to the beam that left the extreme point of the diaphragm at an angle β, should intersect at exactly one point. Therefore, the discriminant of a quadratic equation

$$\frac{z^2}{4k^2x_0^3} = -R + z\,\mathrm{tg}\,\beta \tag{1.63}$$

must be zero:

$$D = \mathrm{tg}^2\,\beta - 4\frac{1}{4k^2x_0^3}R = 0, \tag{1.64}$$

those: $\mathrm{tg}\,\beta = \sqrt{R}\big/\big(kx_0^{3/2}\big)$. The distance itself is equal to the root of this quadratic equation:

$$z_{\max} = \frac{\mathrm{tg}\,\beta}{2\big/\big(4k^2x_0^3\big)} = 2kx_0^{3/2}\sqrt{R},\qquad (1.65)$$

thus, it turned out that the distance coincides with (1.62).

The diffractionless Airy beam is two-dimensional. In the three-dimensional case the diffraction-free Bessel beam is a beam which in practice also retains its invariance at a distance which depends on the radius of the circular diaphragm R. Let us compare expression (1.62) with a similar expression [1] for a limited Bessel beam $J_0(kr\sin\theta)$:

$$z_{\max} = \frac{R}{\mathrm{tg}\,\theta},\qquad (1.66)$$

where r is the radial coordinate, θ is the angle of inclination of the conical wave forming the Bessel beam, $J_0(x)$ is the Bessel function of zero order. Equation (1.66) shows that length of the invariance of the bounded Bessel beam is proportional to the radius of the aperture R, and a similar length of invariance Airy beam (1.62) is proportional to the square root of R. That is, other things being equal, the Bessel beam will remain non-diffracting on a larger portion of the trajectory than the Airy beam.

1.2.2. Getting the invariance distance from the asymptotic behaviour of the Airy beam phase

The Airy function from (1.58) for large (absolute) negative values of the argument $s \ll -1$ has an asymptotics (expression 10.4.60 in [53]), from which it follows that

$$\mathrm{Ai}(-s) \approx \frac{s^{-1/4}}{2i\sqrt{\pi}}\left[\exp\left(i\frac{2}{3}s^{3/2} + \frac{i\pi}{4}\right) - \exp\left(-i\frac{2}{3}s^{3/2} - \frac{i\pi}{4}\right)\right].\qquad (1.67)$$

The field with a complex amplitude (1.67) is a superposition of two fields with opposite phases. The phase of the second field is

$$\phi(s) = -\frac{2}{3}s^{3/2}.\qquad (1.68)$$

The asymptotics (1.68) was used in [41] to form the Airy beam using a liquid-crystal light modulator. We use the expression (1.68) to obtain a formula for calculating the distance of invariance of a bounded Airy beam. The light beam arriving from the initial plane from the point with coordinates $(x = sx_0, z = 0)$ to the point of the Airy beam trajectory with coordinates $(\Delta x, z)$ is determined by the derivative of the phase (1.68)

$$\frac{1}{k}\frac{d}{dx}\varphi(s) = -\frac{x^{1/2}}{kx_0^{3/2}},\qquad(1.69)$$

The derivative of the phase (1.69) is equal to the slope φ of the beam to the z axis:

$$\left|\frac{1}{k}\frac{d}{dx}\varphi(s)\right| = \operatorname{tg}\varphi.\qquad(1.70)$$

On the other hand, the tangent (1.70) is equal to the ratio of the sum of the distance from the origin to the point of the beginning of the beam $x = sx_0$ and the shift Δx (1.60) to the distance z to the observation point on the Airy beam path:

$$\operatorname{tg}\varphi = \frac{x + \Delta x}{z},\qquad(1.71)$$

Equating the right parts (1.69) and (1.71), we get:

$$z \quad 2kx^{3/2}\sqrt{x}.\qquad(1.72)$$

Let the distance x be equal to the absolute value of the coordinates of the end of the aperture bounding the Airy beam $x = R$ in the initial plane, then we obtain the formula for the invariance distance of the bounded Airy beam:

$$z_2 = 2kx_0^{3/2}\sqrt{R}.\qquad(1.73)$$

The first term in (1.67) corresponds to a field whose rays are directed in the opposite direction (symmetrically about the optical axis). These rays intersect only in the region of negative values of z, i.e. form an imaginary caustic, so the contribution from this field does not affect the distance of invariance of the bounded Airy beam.

Comparing (1.62) and (1.73), we conclude that both expressions, obtained in different ways, for the length of the invariance of the Airy bounded on one side are $z_1 = z_2$. In [54], the question of the correspondence of the rays emerging from the initial plane of the Airy beam to the points of the parabolic trajectory of the main maximum was investigated. But the expressions like (1.62) or (1.73) were not obtained in [54].

1.2.3. Simulation results

Figure 1.5 shows the dependence of the maximum intensity of the bounded Airy beam (1.65) on the distance z traversed for different values of diaphragm radius R: 10λ (*a*), 40λ (*b*), 80λ (*c*).

Let us determine the length of the bounded Airy beam as the distance at which the maximum intensity falls to a certain level. To determine this level, consider by analogy of a bounded Bessel beam with a complex amplitude in the initial plane $J_0(r/r_0)\mathrm{circ}(r/R)$ (r is the radial polar coordinate, $\mathrm{circ}(x)$ is a function of the circle, is equal to one at $x \leq 1$ and equal to zero for $x > 1$). When $\lambda = 532$ nm, $R =$

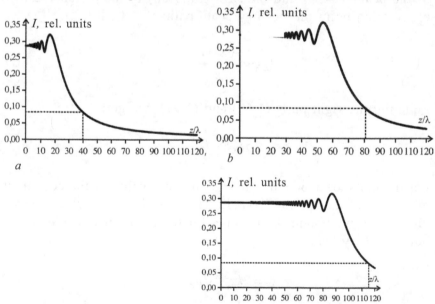

Fig. 1.5. Dependence of he maximum intensity of the boundaed Airy beam (1.60) on the distance z traversed for different values of diaphragm radius R: 10λ (*a*), 40λ (*b*), 80λ (*c*).

10λ and $r_0 = \lambda$ the Bessel beam length according to formula (1.66) is equal to $z_{max} = R[(kr_0)^2-1]^{1/2} \approx 33$ μm. Numerical simulation showed that at this distance the axial intensity of the Bessel beam amounts to 16% of the initial intensity (Fig. 1.6). Figure 1.6 shows that at a distance z_{max} the dependence of the axial intensity on the distance z has no special features. For example, it is not a point of maximum or minimum, and is also not a point of intensity half-decay from the initial level. By analogy, we will conditionally determine the length of the Airy beam so that, for any particular radius of the diaphragm, this length coincides with the value calculated by the formula (1.62). We choose this diaphragm radius equal to $R = 10\lambda$. The length of this bounded Airy beam according to (1.62) is equal to $z_{max} \approx 39.7\lambda$. The intensity of the beam at this distance, calculated by the BPM method, is 29.3% of the initial level. Then for all other diaphragm radii we determine the beam length as the distance at which the intensity drops to 29.3% (Fig. 1.5). Figure 1.7 shows the dependence

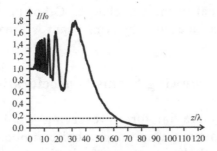

Fig. 1.6. The dependence of the maximum (axial) bounded Bessel beam intensity of the distance traversed z ($R = 10\lambda$). The dotted line indicates the distance $z_{max} = 33$ μm and the intensity $I/I_0 = 0.16$.

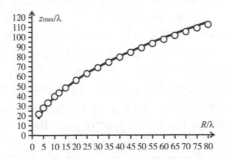

Fig. 1.7. The dependence of the invariance distance of a bounded Airy beam on the radius of the diaphragm, calculated from the formula (1.62) (solid curve) and by the BPM method (dotted curve).

of the invariance distance of the limited Airy beam on the diaphragm radius, calculated by formula (1.62) and the BPM method.

Simulation has shown that, with a small radius of the diaphragm, the theoretical and calculated distances of invariance differ significantly, however, with $R \geq 10\lambda$, the error is less than 5%.

In this section, we obtain an explicit expression that determines the distance of the invariance of the bounded Airy beam, i.e. the distance at which the Airy beam, bounded in the initial plane from the side of a slow decrease in amplitude, propagates along a parabolic trajectory without changing amplitude (almost without diffraction). This distance is obtained in two ways – from the equation of the trajectory and from the distribution of the phase of the Airy beam. Unlike the same distance for the Bessel diffraction-free beam, proportional to the radius of the circular diaphragm bounding the beam, the distance of invariance for the Airy beam is proportional to the square root of the edge coordinate (in absolute value) of the restricting diaphragm. For the Bessel beam, this distance does not depend on the wavelength of the light, while for the Airy beam it is inversely proportional to the wavelength. Other things being equal, the distance of invariance for the Airy beam is greater than for the Bessel beam.

1.3. Conversion of slowing beams to accelerating ones

Recently [55], nonparaxial form-preserving Webber accelerating beams, which propagate along a parabolic trajectory, have been considered. These Weber beams are similar to the 'half Bessel beams' [44], but unlike the latter, they are described by an analytical expression. The Weber–Hermite beams are also known as solutions of the paraxial propagation equation [56]. The general theory of 3D nonparaxial accelerated beams based on the well-known solutions of the Helmholtz equation in parabolic, elongated and oblate spheroidal coordinates is considered in [57]. These beams propagate along a circular arc. In [50] Airy beams are proposed that propagate with non-uniform acceleration along a hyperbolic trajectory. Although these beams do not retain their shape (diverge during propagation), they may have a more curved trajectory at its final part than yjr ordinary Airy beams [7].

In this section, a different approach to the formation of accelerated beams is considered. It consists in the following. Paraxial 2D light fields are known, in which the argument of the complex amplitude function depends on the variables as x^2/z, where x is the transverse coordinate and z is the longitudinal coordinate. This, for example, is the light field, which is formed when a plane wave is diffracted by an angular phase step [58]. Or the well-known solution to the problem of diffraction on the edge of an opaque screen [59].

In this section, we will consider other solutions of the paraxial propagation equation. Light fields whose complex amplitude has an argument of the form x^2/z, propagate along the root path of the parabola $x = z^{1/2}$. Such beams are slowed down, since 'acceleration' (the second derivative along the trajectory) $x'' = -z^{3/2}$ has the opposite sign with speed (the first derivative along the trajectory) $x' = z^{-1/2}$. If the amplitude of such a light field at a distance z_0 is replaced by the complex conjugate and shift the beginning of the optical axis to the point z_0, then the light field with such an amplitude will propagate with acceleration along the trajectory $x = (z_0 - z)^{1/2}$. In this work, analytical expressions are given for the complex amplitudes of such accelerating beams. In addition, the paraxial beams of the 'Bessel half', which differ from the nonpaxial ones, are given [44].

1.3.1. Accelerating beams

Let us assume that for each fixed distance travelled z the coordinate of the maximum intensity of a certain laser beam has the form $x_{max}(z)$. In order for the beam trajectory to have acceleration at a certain section, it is necessary that the first and second derivative coordinates x_{max} of the maximum over the travelled distance z have the same sign [50]:

$$\left(\frac{dx_{max}}{dz}\right)\left(\frac{d^2 x_{max}}{dz^2}\right) > 0. \tag{1.74}$$

The most widely known accelerating beams are the Airy beams, whose complex amplitude has the form [7]:

$$E(x,z) = Ai\left(s - \xi^2/4\right)\exp\left(is\xi/2 - i\xi^3/12\right), \tag{1.75}$$

wherein (x,z) are the Cartesian coordinates, $s = x/x_0$, $\xi = z/(kx_0^2)$, $k = 2\pi/\lambda$, is the wave number, λ is the wavelength, x_0 is an arbitrary scaling factor, $Ai(x)$ is the Airy function [53, section 10.4]. The coordinates of the intensity maxima of such beams have the form:

$$x_{max} = x_0 y_m + \frac{z^2}{4k^2 x_0^3}, \qquad (1.76)$$

where y_m is the point of the m-th maximum of the function $[Ai(x)]^2$.
Then

$$dx_{max}/dz = z/(2k^2 x_0^3) \quad \text{and} \quad d^2 x_{max}/dz^2 = 1/(2k^2 x_0^3).$$

That is, condition (1.73) is satisfied for any distances $z > 0$, and the
acceleration $d^2 x_{max}/dz^2$ has a constant value. Below we consider laser
beams, which also have acceleration, but which is not constant and
decreases as the beam propagates.

Airy beams with hyperbolic trajectory

In [50], the Airy beams with a hyperbolic trajectory, which in the
initial plane $z = 0$ have a complex amplitude

$$E(x,0) = \exp\left[i\alpha (x/x_0)^3 + i\beta (x/x_0) \right], \qquad (1.77)$$

where x_0 is the scaling factor, a α and β are dimensionless parameters,
are considered. The trajectory of such a beam in the Fresnel
diffraction zone has the form:

$$x_{max} = \frac{\left(\beta - y_m \sqrt[3]{3\alpha}\right)z}{kx_0} - \frac{kx_0^3}{12\alpha z}. \qquad (1.78)$$

Using the condition (1.73), it was shown in [50] that acceleration
occurs at

$$z > z_1 = \frac{kx_0^2}{2\sqrt{3\alpha\left(y_m \sqrt[3]{3\alpha} - \beta\right)}}, \qquad (1.79)$$

and only in the case when sign $(\alpha)\beta < y_m(3|\alpha|)^{1/3}$. Acceleration of
the beam decreases proportionally to z^{-3}: $d^2 x_{max}/dz^2 = -kx_0^3/(6\alpha z^3)$.

Hermite–Gaussian beams

The widely known Hermite–Gaussian beams [60], it turns out, also have acceleration. Indeed, let the light field in the initial plane $z = 0$ have a complex amplitude

$$E(x, z = 0) = \exp\left(-\frac{x^2}{w^2}\right) H_n\left(\frac{x}{a}\right),\qquad(1.80)$$

where (x,z) are the Cartesian coordinates, w is the radius of the Gaussian beam waist, n and a is the order and scale of the Hermite polynomial. Then, applying the Fresnel transform, it can be shown that at a distance z from the initial plane a field will be formed with the following distribution of the complex amplitude [61]:

$$E(x,z) = \sqrt{\frac{-ik}{2pz}}\exp\left[-\frac{k}{2z}\left(\frac{k}{2zp}-i\right)x^2\right]\left(1-\frac{1}{pa^2}\right)^{n/2}\times$$

$$\times H_n\left[\frac{-ikx}{2z\sqrt{(pa)^2-p}}\right],\qquad(1.81)$$

where $p = 1/w^2 - ik/(2z)$.

Consider for simplicity the case when $n = 1$. Then the intensity of the beam (1.81) in the plane located at a distance z from the initial plane, is equal to:

$$I(x,z) = |E(x,z)|^2 = \frac{k^3 x^2}{2a^2 |p|^3 z^3}\exp\left(-\frac{k^2 \operatorname{Re} p}{2z^2 |p|^2}x^2\right).\qquad(1.82)$$

Differentiating both sides of (1.81) with respect to a variable x, we obtain the necessary condition for intensity extremes:

$$2x = \frac{k^2 \operatorname{Re} p}{z^2 |p|^2}x^3.\qquad(1.83)$$

The case $x = 0$ corresponds to the minimum (since the intensity $I(0,z)$ is zero), and the coordinates of the maxima are equal to

$$x_{\max} = \pm \sqrt{\frac{z^2 + z_0^2}{kz_0}}, \qquad (1.84)$$

where $z_0 = kw^2/2$ is the Rayleigh distance.

It is easy to show that the curve with maximum intensity is a hyperbole. We obtain the derivatives of $x_{\max}$ the z first and second orders:

$$\frac{dx_{\max}}{dz} = \pm \frac{z}{\sqrt{kz_0}\sqrt{z^2 + z_0^2}},$$

$$\frac{d^2 x_{\max}}{dz^2} = \pm \frac{1}{\sqrt{kz_0}} \frac{z_0^2}{\left(z^2 + z_0^2\right)^{3/2}}. \qquad (1.85)$$

From (1.85) it is clear that for all $z > 0$ the product $(dx_{\max}/dz) \times (d^2 x_{\max}/dz^2)$ positive, i.e. the Hermite–Gaussian beam has acceleration, which, like the Airy beams with a hyperbolic trajectory discussed above, decreases cubically with the distance z. The presence of acceleration in both branches of the first-order Hermite–Gaussian beam may be noticeable at small distances z. Figure 1.8 shows the intensity of such a beam in the plane Oxz, calculated by the beam propagation method (BPM). The two local maxima of the Hermite–Gaussian beam in Fig. 1.8 propagate symmetrically about the optical axis along two hyperbolic trajectories with acceleration.

Note that as the number n of the Hermite–Gaussian beam increases (at $a = w/\sqrt{2}$), the radius (width) of the beam increases and the coordinate of the extreme intensity zeros (at different distances z) is described by an upper estimate:

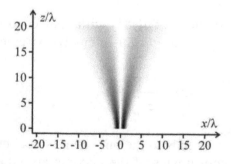

Fig. 1.8. Accelerating Hermite–Gaussianian beam ($\lambda = 532$ nm, $w = \lambda$, $a = 2\lambda$, $n = 1$).

$$\left| x_{max}^{n} \right| \leq \frac{w}{2} \sqrt{n(n-1)} \sqrt{1+\frac{z^2}{z_0^2}}, \tag{1.86}$$

from which it follows, in view of (1.84), that the magnitude of the Hermite–Gaussian beam acceleration with an increase in the number n grows linearly for large n.

Other paraxial 2D accelerated beams, which are described analytically, are not yet known. Therefore, the following section shows how to accelerate laser beams from slower ones.

1.3.2. Slowing beams

In contrast to the accelerating beams (1.74), (1.76), (1.79) considered above, for the slowing beams it is necessary that the first and second derivative coordinates of the maximum intensity x_{max} over the distance z travelled be of a different sign:

$$\left(\frac{dx_{max}}{dz} \right) \left(\frac{d^2 x_{max}}{dz^2} \right) < 0. \tag{1.87}$$

Consider the examples of such beams.

Diffraction of a plane wave on an opaque semi-infinite screen
Let a plane wave propagate along the optical axis z and when $z = 0$ it passes through a semi-infinite flat aperture that transmits light in the region $x < 0$ (x is the coordinate in the plane of the aperture). Directly behind it, the complex amplitude will be equal to:

$$E(x, z = 0) = \begin{cases} 1, x < 0, \\ 0, x \geq 0. \end{cases} \tag{1.88}$$

After the light passes the distance z, its complex amplitude will be determined by the Fresnel transformation from the field (1.88):

$$E(x, z) = \frac{1}{2} \left\{ \left[1 - C(\xi) - S(\xi) \right] + i \left[C(\xi) - S(\xi) \right] \right\}, \tag{1.89}$$

where $\xi = \sqrt{2/(\lambda z)} x$, and $C(\xi)$ and $S(\xi)$ is the Fresnel integral:

$$C(\xi) = \int\limits_0^\xi \cos\left(\frac{\pi t^2}{2}\right) dt, \quad S(\xi) = \int\limits_0^\xi \sin\left(\frac{\pi t^2}{2}\right) dt. \tag{1.90}$$

The coordinates of the intensity maxima of such a beam x_{max} are equal to:

$$x_{max} = \sqrt{\frac{\lambda z}{2}} \xi_m, \tag{1.91}$$

where ξ_m is the coordinate of the m-th maximum of the function $I(\xi) = [1 - C(\xi) - S(\xi)]^2 + [C(\xi) - S(\xi)]^2$. Calculate the first and second derivative coordinates of the maximum x_{max} of the distance travelled z:

$$\frac{dx_{max}}{dz} = \frac{1}{2}\sqrt{\frac{\lambda}{2z}}\xi_m,$$

$$\frac{d^2 x_{max}}{dz^2} = -\frac{1}{4z}\sqrt{\frac{\lambda}{2z}}\xi_m. \tag{1.92}$$

From (1.92) it is clear that condition $z > 0$ (1.87) is satisfied for all. The slowdown of each maximum can be seen in Fig. 1.9, which shows the beam intensity (1.89).

Two-dimensional hypergeometric beams and Bessel beams
Just as it was done in [62], we will seek a solution to the paraxial propagation equation

$$2ik\frac{\partial E}{\partial z} + \frac{\partial^2 E}{\partial x^2} = 0, \tag{1.93}$$

in the form $E(x,z) = x^p z^q F(sx^m z^n)$, where $F(x)$ is a certain function, s is a scaling factor. Reducing the obtained second-order differential equation to the Kummer equation, we obtain the following expression for the complex amplitude of the 2D analogue of the 3D generalized hypergeometric mode [63]:

$$E(x,z) = z^{-a} {}_1F_1\left(a, \frac{1}{2}, \frac{ikx^2}{2z}\right), \tag{1.94}$$

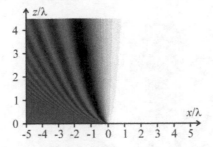

Fig. 1.9. The intensity in the plane Oxz in the diffraction plane wave on a semi-infinite aperture ($\lambda = 532$ nm, $-5\lambda \leq x \leq 5\lambda$, $0.1\lambda \leq z \leq 5\lambda$).

with a is an arbitrary constant and $_1F_1$ is the Kummer function. It is known that any solution of the derivative (1.93) of any of the Cartesian coordinate as a solution of equation (1.93). Therefore, we can consider a light beam with an amplitude

$$E(x,z) = xz^{-a} \, _1F_1\left(a, \frac{3}{2}, \frac{ikx^2}{2z}\right). \tag{1.95}$$

In the particular case when $a = 3/4$ in (1.95) follows a solution of (1.93) as a Bessel beam fractional order:

$$E(x,z) = \sqrt{\frac{x}{z+z_0}} J_{\frac{1}{4}}\left[\frac{kx^2}{4(z+z_0)}\right] \exp\left[\frac{ikx^2}{4(z+z_0)}\right], \tag{1.96}$$

where z_0 is an arbitrary positive constant (so that there is no singularity in the plane $z = 0$). When obtaining (1.95), identity 13.6.1 from [53] was used. The coordinates of the intensity maxima of such a beam have the form:

$$x_{\max} = \sqrt{\frac{4(z+z_0)}{k}} y_m, \tag{1.97}$$

where y_m is the m-th root of the equation $J_{1/4}(y)\left[J_{1/4}(y) + 4J'_{1/4}(y)y\right] = 0$. Dependence (1.97) is similar to dependence (1.91), i.e. $x_{\max}$ is proportional to $z^{1/2}$. Therefore, the beam (1.96) is slowing down, as can be seen from Fig. 1.10, which shows the intensity of such a beam, calculated by the BPM method ($\lambda = 532$ nm, $z_0 = 20\lambda$, modelling domain $-20\lambda \leq x \leq 20\lambda$, $0 \leq z \leq 80\lambda$).

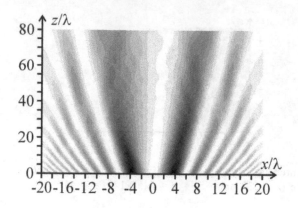

Fig. 1.10. Intensity in the plane Oxz of the light beam (1.96).

1.3.3. Slowing and accelerating beams in different parts of the trajectory

From (1.73) and (1.87) it is easy to notice that a simple change of variables $z \to z_0 - z$ leads to a change in acceleration to deceleration and vice versa. In fact, consider a light beam, the complex amplitude of which is obtained from (1.89) by the complex conjugation and replacing ζ with $\xi = \sqrt{2/\left[\lambda(z_0 - z)\right]} x$:

$$E(x, z < z_0) = \frac{1}{2} \left\{ \begin{matrix} [1 - C(\xi) - S(\xi)] - \\ -i[C(\xi) - S(\xi)] \end{matrix} \right\}. \tag{1.98}$$

The light beam (1.98) will be called the Fresnel beam. Using limits

$$\lim_{x \to \infty} C(x) = \frac{1}{2}, \lim_{x \to \infty} S(x) = \frac{1}{2}, \tag{1.99}$$

we obtain that near the plane $z = z_0$ (near the focal plane), but when $z < z_0$, the complex amplitude has the form:

$$E(x, z \to z_0 - 0) = \begin{cases} 0, x > 0, \\ 1, x < 0. \end{cases} \tag{1.100}$$

The intensity of the light beam (1.98) in the plane Oxz is shown in Fig. 1.11.

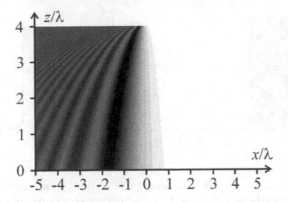

Fig. 1.11. The intensity of the light beam (1.98) in the plane Oxz (wavelength $\lambda = 532$ nm, the distance from $z = 0$ the focal plane is equal to $z_0 = 4.1\lambda$. Modelling domain: $-5\lambda \leq x \leq 5\lambda$, $0 \leq z \leq 4\lambda$).

Directly behind the plane $z = z_0$, the arguments ξ of the Fresnel integrals in (1.98) become imaginary, since $2/[\lambda(z_0-z)] < 0$, in addition, the imaginary part can be both positive and negative, therefore two solutions are possible, of which only one satisfies the boundary condition $E(x,z \to z_0+0) = E(x,z \to z_0-0)$. Using identities for the Fresnel integrals of imaginary variables $C(iz) = iC(z)$ and $S(iz) = -iS(z)$ [53, expressions 7.3.18], it can be shown that behind the plane $z = z_0$ the complex amplitude has the form:

$$E(x,z > z_0) = \frac{1}{2}\left\{ \begin{array}{l} \left[1-C(\eta)-S(\eta)\right]+ \\ +i\left[C(\eta)-S(\eta)\right] \end{array} \right\},\tag{1.101}$$

where $\eta = \sqrt{2/\left[\lambda(z-z_0)\right]}\,x$. The second solution is equal to

$$E(x,z > z_0) = \frac{1}{2}\left\{ \begin{array}{l} \left[1+C(\eta)+S(\eta)\right]- \\ -i\left[C(\eta)-S(\eta)\right] \end{array} \right\},\tag{1.102}$$

does not satisfy the boundary condition.

From comparison of (1.98) and (1.101) it can be seen that the light beam is accelerating when $z < z_0$, focussing when $z = z_0$ in a uniform semi-infinite spot, and the amplitude at $z > z_0$ is a mirror reflection of the amplitude at $z < z_0$, i.e. distributed with a slowdown. Figure 1.12 *a* shows the intensity of the light beam (1.98), (1.101) in the plane Oxz, and Fig. 1.12*b* shows the intensity cross

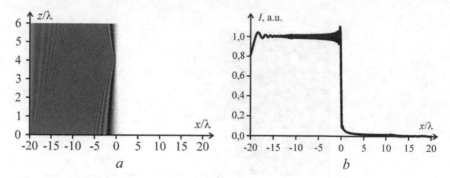

Fig. 1.12. The intensity of the light beam (1.98), (1.101), calculated by the FDTD method: (*a*) intensity in the plane *Oxz* and (*b*) cross-section of intensity in the plane $z = z_0$.

section in the plane $z = z_0$. Figure 1.12 was obtained by calculating by the BPM method at $\lambda = 532$ nm, $z_0 = 4\lambda$, modelling domain $-20\lambda \leq x \leq 20\lambda$, $0 \leq z \leq 8\lambda$. Intensity oscillations in the vicinity $x = 0$ are explained by the limited initial modelling domain.

Similarly, replacing in (1.96) $z + z_0$ with $z_0 - z$ and applying complex conjugation, we obtain a light beam, whose complex amplitude in the initial plane ($z = 0$) is equal to

$$E(x, z = 0) = \sqrt{\frac{x}{z_0}} J_{\frac{1}{4}}\left(\frac{kx^2}{4z_0}\right) \exp\left(-\frac{ikx^2}{4z_0}\right). \qquad (1.103)$$

By setting the initial field (1.103), using the BPM method, we calculated the intensity in the plane *Oxz* of the light beam (at $\lambda = 532$ nm, $z_0 = 4\lambda$), which, as can be seen from Fig. 1.13, focuses with acceleration. The asymmetry of the focal spot arising in this case is explained by the presence of a phase jump on $\pi/2$ in the field (1.102) at the point $x = 0$. Such focussing with acceleration was considered earlier in [64] for a radially symmetric Airy beam.

Equate to zero the amplitude of the field in the initial plane with $x < 0$:

$$E(x, z = 0) = \begin{cases} \sqrt{\frac{x}{z_0}} J_{\frac{1}{4}}\left(\frac{kx^2}{4z_0}\right) \exp\left(-\frac{ikx^2}{4z_0}\right), & x \geq 0, \\ 0, & x < 0. \end{cases} \qquad (1.104)$$

Figure 1.14 shows the beam intensity (1.104) obtained by the BPM method with the same parameters as in Fig. 1.102.

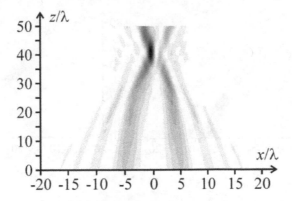

Fig. 1.13. Intensity in the plane *Oxz* of the accelerating light beam with the distribution of the complex amplitude in the initial plane (1.103).

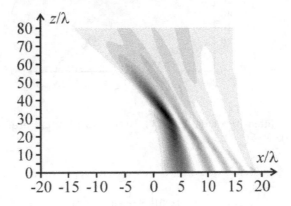

Fig. 1.14. Intensity in the plane *Oxz* of the accelerating light beam with the distribution of the complex amplitude in the initial plane (1.104).

For other orders of the Bessel function in (1.104), the picture type is similar (Fig. 1.15).

From Fig. 1.15 it can be seen that the acceleration decreases with increasing order of the Bessel function. The accelerating paraxial beams (1.104) are similar to the nonparaxial beams of the 'half of Bessel' [44], therefore we will call them the paraxial beams of the 'half of Bessel'.

Diffraction of a Gaussian beam on a semi-infinite opaque screen
Let us consider another example, when a slowing beam is described by an analytic function. Let a Gaussian beam with a waist radius *w* pass through a semi-infinite flat aperture. Directly behind the aperture, the complex amplitude will be equal to:

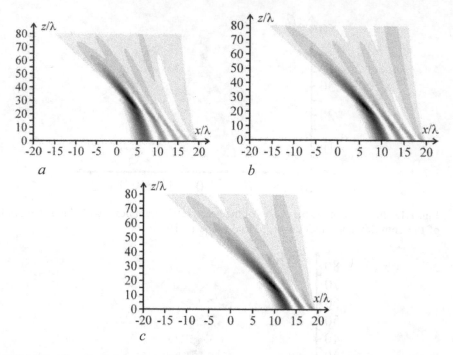

Fig. 1.15. The intensity of the beam (1.104) with different orders of the Bessel function: 1 (*a*), 3 (*b*) and 5 (*c*).

$$E(x, z = 0) = \begin{cases} \exp\left(-\dfrac{x^2}{w^2}\right), & x < 0, \\ 0, & x \geq 0. \end{cases} \qquad (1.105)$$

After the light passes the distance z, its complex amplitude will be determined by the Fresnel transformation from the field (1.105) and takes the form [65,66]:

$$E(x, z) = \sqrt{\frac{-ik}{8pz}} \exp\left[\frac{ikx^2}{2(z - iz_0)}\right] \text{erfc}\left(\frac{-ikx}{2\sqrt{pz}}\right), \qquad (1.106)$$

where $p = 1/w^2 - ik/(2z)$, $z_0 = kw^2/2$ is the Rayleigh distance, erfc(x) is an additional error function:

$$\text{erfc}(z) = 1 - \text{erf}(z) = 1 - \frac{2}{\sqrt{\pi}} \int_0^z \exp(-t^2) dt. \qquad (1.107)$$

The equation (1.106) can be reduced to the form:

$$E(x,z) = \sqrt{\frac{-ik}{8pz}} \exp\left(\frac{ikx^2}{2z}\right) \exp(-y^2) \operatorname{erfc}(iy), \qquad (1.108)$$

where

$$y = \frac{-kx}{2\sqrt{pz}} = -\left(\frac{x}{w}\right)\left(\frac{z_0}{z}\right)\left(1 + \frac{z_0^2}{z^2}\right)^{-1/4} \exp\left[i\frac{1}{2}\arctan\left(\frac{z_0}{z}\right)\right]. \qquad (1.109)$$

When $z \to 0$ the argument variable y is almost independent of z: $\arg(y) \approx \pi/4$. Therefore, the equation of the trajectory of such a beam at small distances z has the form:

$$x_{max} = -w\eta_m\sqrt{\frac{2z}{z_0}}, \qquad (1.110)$$

where η_m is the m-th maximum of the function $|\operatorname{erfc}[\eta(i-1)]|^2$. This means that, like the light beam (1.89), the beam (1.109) will slow down.

By analogy with the formation of a uniform intensity distribution on the half-plane (using a beam (1.98), (1.101)), we will create a distribution (1.105) in the plane $z = z_0$ using a light beam with the following distribution of complex amplitudes:

$$E(x,z) = \sqrt{\frac{ik}{8p(z_0 - z)}} \exp\left[\frac{-ikx^2}{2(z_0 - z + iz_0)}\right] \operatorname{erfc}\left[\frac{ikx}{2\sqrt{p}(z_0 - z)}\right], \qquad (1.111)$$

where $p = 1/w^2 + ik/[2(z_0-z)]$. In (1.111), the probability integral or the Laplace integral is included as a factor. Therefore, we will refer to beams (1.111) as Laplace beams. From the form of the trajectory (1.110), it follows that the beam (1.111) will also be accelerating near the plane $z = z_0$. This can be seen in Fig. 1.16 *a*, which shows the beam intensity (1.111) in the plane Oxz. Figure 1.16 *b* shows the intensity cross section in the plane $z = z_0$.

The following results are obtained in the section. It is shown that the well-known Hermite–Gaussian modes and the generalized Hermite–Gaussian beams (1.80) are accelerating beams, that is, two extreme local maxima of the intensity, symmetric about the optical axis, propagate along hyperbolic trajectories with non-uniform acceleration, which decreases proportionally to the cube of distance (1.84). A method is proposed for converting two-dimensional light beams propagating with deceleration into accelerating light beams. The Fresnel beam (1.98) accelerated at the final part of the trajectory (1.98), which is obtained from a complex amplitude describing the diffraction of a plane wave on a non-transparent screen (1.89), by complex conjugation and shift along the optical axis. A paraxial accelerated on a finite segment beam of the 'half Bessel' type (1.104), which propagates along the root path of the parabola, and obtained from a generalized hypergeometric laser beam (1.95), by complex conjugation, shift along the optical axis and 'half capture' is considered. The Laplace beam (1.111) accelerated on a finite segment, obtained on the basis of solution (1.15) of the problem of diffraction of a Gaussian beam on an opaque screen, by complex conjugation and shift along the optical axis of the complex amplitude (1.106) is considered in a similar way.

1.4. Beams with a curved trajectory in a linear gradient medium

Integral transformations are often used to describe the propagation of light fields in homogeneous media and various optical systems. The most universal are the Stretton–Chu formulas [67], which make it possible to determine the complex amplitude of light at any point if the complex amplitude is known on an arbitrary surface (without currents and charges). Many other less universal integral transformations can be obtained from the Stretton–Chu formulas. Thus, in a homogeneous medium, light is described by the Rayleigh–Sommerfeld transform

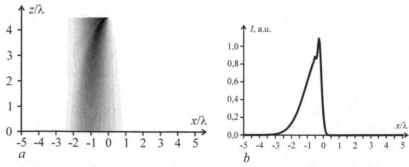

Fig. 1.16. The intensity of the light beam (1.111) in the plane Oxz (a), as well as the intensity cross section in the plane $z = z_0$ (b).

[68, 69]. If the distance from the initial plane to the observation plane is much larger than the wavelength, then this transformation can be replaced by the Kirchhoff transformation [67], which, with the paraxial propagation of light, becomes the Fresnel transformation [67]. If the light propagates not in a homogeneous 2D environment, but through the optical system described by the ABCD matrix, then, with paraxial propagation, the complex amplitudes in the input and output planes are interconnected by the ABCD transformation [2]:

$$E(x,z) = \sqrt{\frac{-ik}{2\pi B}} \int_{-\infty}^{+\infty} E(\xi,0) \exp\left[\frac{ik}{2B}\left(A\xi^2 - 2x\xi + Dx^2\right)\right] d\xi. \quad (1.112)$$

where ξ and x are the transverse Cartesian coordinates in the input and output planes, respectively, $k = 2\pi/\lambda$ is the wave number, λ is the wavelength of light. In particular, if the optical system is a thin lens, then the ABCD transform takes the form of a Fourier transform. The ABCD transformation caused an increased interest due to its versatility, since it could be applied to any systems described by the ABCD matrix. Many papers are devoted to the effective calculation of integrals with a complex exponent, which have a quadratic phase [70, 71]. Later it turned out that such integral transformations also arise in gradient media. The ABCD transform describes the propagation of light in a gradient waveguide with a parabolic dependence of the refractive index on the transverse coordinate [72–74, 22]:

$$n(x) = n_0\left(1 - \frac{x^2}{2a^2}\right).$$

In this case, the ABCD matrix has the form:

$$\begin{pmatrix} A & B \\ C & D \end{pmatrix} = \begin{pmatrix} \cos(z/a) & a\sin(z/a) \\ -\sin(z/a)/a & \cos(z/a) \end{pmatrix}. \quad (1.113)$$

where n_0 is the refractive index on the axis of the waveguide ($x = 0$), a is the index characterizing the rate of decline of the refractive index from the axis of the waveguide to its edge, z is the distance between the input and output planes.

The solution of the paraxial propagation equation for a planar inhomogeneous along the optical axis of a gradient medium with a linear profile in the form of a Fourier transform of an unknown spatial spectrum was obtained in [37].

In this section, based on the methodology of [37], it is shown that in a 2D gradient medium with a linear dependence of the dielectric constant on the transverse coordinate $n^2(x) = n_0^2(1 - \alpha x)$ in the paraxial approximation, the light propagation is described by an integral transform whose kernel includes a complex exponent with a quadratic phase. The propagation of light at a distance z in such a medium is equivalent to passing through a prism whose strength is proportional to z, subsequent propagation in a homogeneous medium with a refractive index n_0, repeated passage through the said prism, and an additional constant phase shift depending on the cubic distance z. It is shown that when a Gaussian beam propagates, its centre shifts along a parabola, proportional to z^2, and its radius coincides with the radius of a Gaussian beam propagating in a homogeneous medium with a refractive index n_0. A Gaussian beam accelerating along a parabola is similar to an Airy beam, also accelerating along a parabola in a 2D homogeneous space [17, 18]. Using the obtained integral transform, an explicit expression is found for the complex amplitude of the Airy beam in a planar gradient medium with a linear profile. This solution coincides with that obtained earlier in [71]. But work [71] does no described the method of obtaining solutions. It is shown that when the Airy beam is matched with the gradient of a linear medium, the Airy beam propagates along a straight-line trajectory.

1.4.1. Integral transform for linear gradient media

Let a 2D gradient medium with a linear dependence of the dielectric constant (square of the refractive index) on the transverse Cartesian coordinate be given:

$$n^2(x) = n_0^2(1 - \alpha x), \qquad (1.114)$$

where x is the Cartesian coordinate in the plane perpendicular to the optical axis z, n_0 is the refractive index on the optical axis, α is the parameter for changing the dielectric constant with distance from the optical axis in directions along the x coordinates. With TE-polarization, the Helmholtz equation for the complex amplitude of light E_y propagating in such a medium, has the form:

$$\frac{\partial^2 E}{\partial x^2} + \frac{\partial^2 E}{\partial z^2} + k_0^2 n_0^2 (1 - \alpha x) E = 0, \tag{1.115}$$

where $k_0 = 2\pi/\lambda_0$ is the wave number in vacuum, λ_0 is the wavelength of light in vacuum. We assume that the light propagates mainly in the direction of the optical axis and represent the amplitude in the form:

$$E(x,z) = U(x,z) \exp(ikz), \tag{1.116}$$

where $k = k_0 n_0$ is the wave number on the optical axis. Then, neglecting the second derivative with respect to z, we obtain the following equation from the Helmholtz equation:

$$2ik \frac{\partial U}{\partial z} + \frac{\partial^2 U}{\partial x^2} - k^2 \alpha x U = 0, \tag{1.117}$$

whose solution we will look for in the form:

$$U(x,z) = A \int_{-\infty}^{+\infty} S(u) \exp\left[i \left(Bx^2 + Cu^2 + Dux + Ex + Fu \right) \right] du, \tag{1.118}$$

where A, B, C, D, E, F are functions of z, and S is an arbitrary function.

We substitute this expression into the paraxial Helmholtz equation (1.117), we give similar terms and, taking into account the arbitrariness of the function $S(u)$, we obtain:

$$-A\left[2k\frac{dB}{dz} + 4B^2 \right]x^2 - A\left[2k\frac{dC}{dz} + D^2 \right]u^2 - A\left[2k\frac{dD}{dz} + 4BD \right]ux -$$

$$-A\left[2k\frac{dE}{dz} + 4BE + k^2\alpha \right]x - A\left[2k\frac{dF}{dz} + 2DE \right]u + \tag{1.119}$$

$$+\left[2ik\frac{dA}{dz} + 2iAB - AE^2 \right] = 0.$$

Since this expression is true for any z, we equate the coefficients with similar terms to zero. Get a system of six ordinary differential equations of the first order:

$$\begin{cases} k\dfrac{dB}{dz}+2B^2=0, \\[2mm] 2k\dfrac{dC}{dz}+D^2=0, \\[2mm] k\dfrac{dD}{dz}+2BD=0, \\[2mm] 2k\dfrac{dE}{dz}+4BE+k^2\alpha=0, \\[2mm] k\dfrac{dF}{dz}+DE=0, \\[2mm] 2ik\dfrac{dA}{dz}+2iAB-AE^2=0. \end{cases} \qquad (1.120)$$

The solution of the first equation are functions of the form:

$$B(z)=\frac{k}{2(z+B_0)}, \qquad (1.121)$$

where B_0 is an arbitrary constant.

Substituting (1.121) into the third equation (1.120) and solving it, we get:

$$D(z)=\frac{D_0}{z+B_0}, \qquad (1.122)$$

where D_0 is an arbitrary constant.

Substituting (1.122) into the second equation of the system (1.120), we obtain the form of the functions $C(z)$:

$$C(z)=\frac{D_0^2}{2k(z+B_0)}+C_0, \qquad (1.123)$$

where C_0 is an arbitrary constant.

Substituting (1.121) into the fourth equation of system (1.120) and solving it, we obtain the form of the functions $E(z)$:

$$E(z)=\frac{E_0}{z+B_0}-\frac{k\alpha}{4}(z+B_0), \qquad (1.124)$$

where E_0 is an arbitrary constant.

Substituting the obtained expressions for the functions $D(z)$ and $E(z)$ into the fifth equation of the system (1.120) and solving it, we obtain the form of the functions $F(z)$:

$$F(z) = \frac{\alpha D_0 (z + B_0)}{4} + \frac{D_0 E_0}{k(z + B_0)} + F_0, \qquad (1.125)$$

where F_0 is an arbitrary constant.

The sixth equation of the system (1.120) is solved in separable variables. Substitute the found expressions for the other functions, we get the solution:

$$A(z) = \frac{A_0}{\sqrt{z + B_0}} \exp\left[\frac{iE_0^2}{2k(z + B_0)}\right] \exp\left[-\frac{ik\alpha^2}{96}(z + B_0)^3 + \frac{i\alpha E_0 z}{4}\right]. \quad (1.126)$$

The amplitude in the initial plane $z = 0$ has the form:

$$U(x,0) = A^0 \exp\left[i\left(B^0 x^2 + E^0 x\right)\right] \times$$
$$\times \int_{-\infty}^{+\infty} S(u) \exp\left[i\left(C^0 u^2 + D^0 ux + F^0 u\right)\right] du, \qquad (1.127)$$

where $A^0 = A(0)$, $B^0 = B(0)$, $C^0 = C(0)$, $D^0 = D(0)$, $E^0 = E(0)$, $F^0 = F(0)$.

Multiply the initial field by a factor $\exp[-iB^0 x^2 - i(E^0 + D^0 \xi)x]$ and integrating over the entire numerical axis, we get:

$$\int_{-\infty}^{+\infty} U(x,0) \exp\left[-iB^0 x^2 - i\left(E^0 + D^0 \xi\right)x\right] dx =$$
$$= A^0 \int_{-\infty}^{+\infty} S(u) \exp\left[i\left(C^0 u^2 + F^0 u\right)\right] \left\{\int_{-\infty}^{+\infty} \exp\left[iD^0 (u - \xi)x\right] dx\right\} du. \qquad (1.128)$$

The internal integral is the Dirac delta function and therefore

$$\int_{-\infty}^{+\infty} U(x,0) \exp\left[-iB^0 x^2 - i\left(E^0 + D^0 \xi\right)x\right] dx =$$
$$= \frac{2\pi A^0}{D^0} S(\xi) \exp\left[i\left(C^0 \xi^2 + F^0 \xi\right)\right], \qquad (1.129)$$

thus, the function $S(u)$ is expressed in terms of the initial field as follows:

$$S(u) = \frac{D^0}{2\pi A^0} \exp\left[-i\left(C^0 u^2 + F^0 u\right)\right] =$$

$$= \int_{-\infty}^{+\infty} U(\xi, 0) \exp\left[-iB^0 \xi^2 - i\left(E^0 + D^0 u\right)\xi\right] d\xi. \tag{1.130}$$

Substituting (1.120) into (1.118) and integrating the emerging Poisson integral over u, we obtain:

$$U(x,z) = \frac{D^0 A}{2\pi A^0} \exp\left[i\left(Bx^2 + Ex\right)\right] \int_{-\infty}^{+\infty} U(\xi, 0) \exp\left[-i\left(B^0 \xi^2 + E^0 \xi\right)\right] \times$$

$$\times \sqrt{\frac{i\pi}{C - C^0}} \exp\left[\frac{\left(Dx - D^0 \xi + F - F^0\right)^2}{4i\left(C - C^0\right)}\right] d\xi. \tag{1.131}$$

Substituting the functions (1.121)–(1.126) into (1.131), we obtain the integral transformation connecting the complex amplitudes of light in two planes transverse to the optical axis:

$$U(x,z) = \sqrt{\frac{-ik}{2\pi z}} \exp\left(-\frac{ik\alpha^2 z^3}{96}\right) -$$

$$- \int_{-\infty}^{+\infty} U(\xi, 0) \exp\left[\frac{ik}{2z}(\xi - x)^2 - \frac{ik\alpha z}{4}(x + \xi)\right] d\xi. \tag{1.132}$$

It is easy to see that when $\alpha = 0$, the integral transform (1.132) transforms into the well-known Fresnel transform.

In the three-dimensional case in a medium with a linear dependence of the dielectric constant on the transverse Cartesian coordinates $n^2(x,y) = n_0^2(1 - \alpha x - \beta y)$, a similar integral transform is obtained by multiplying the transformations (1.132) in both coordinates.

From (1.132) it can be seen that the propagation of light at a distance z in the medium (1.114) is equivalent to passing through a prism, the force of which is proportional to z, subsequent propagation in a homogeneous medium with a refractive index n_0, repeated passage through the said prism, and an additional constant phase shift depending cubically on the distance covered z.

The transformation (1.132) is not a convolution, although it can be calculated using the Fourier transform. The transformation (1.132) does not describe the propagation of light in the ABCD system (1.112), but mathematically it does not differ from it, since the kernel is also an exponent with an indicator in a quadratic form. Consequently, the well-known solutions for a homogeneous medium (Gauss–Hermite beams, Airy beams, etc.) can be generalized for the medium (1.114), and the well-known fast methods for calculating integrals with a complex exponent having a quadratic phase [7, 75] are also suitable for modelling light beams in the medium (1.141).

Note that the ABCD transform (1.112) with the matrix (1.113) for a waveguide with a parabolic distribution of the refractive index is not really a solution to the paraxial Helmholtz equation and is its approximate solution only for small changes in the refractive index. At the same time, the transformation (1.132) is an exact solution of equation (1.117).

1.4.2. Propagation of a Gaussian beam in a two-dimensional linear gradient medium

For example, consider the propagation of a two-dimensional Gaussian beam with a waist radius w:

$$U(\xi,0) = \exp\left(-\frac{\xi^2}{w^2}\right). \tag{1.133}$$

Then at a distance z from the waist the beam will have the following complex amplitude:

$$U(x,z) = \left\{\frac{w_0}{w(z)}\exp\left[i\zeta(z)\right]\right\}^{\frac{1}{2}}$$

$$\exp\left\{-\frac{\left[x-x_0(z)\right]^2}{w^2(z)} + \frac{ik\left[x-x_1(z)\right]^2}{2R(z)} + i\Phi(z)\right\}. \tag{1.134}$$

wherein $z_R = kw^2/2$ is the Rayleigh distance, $w(z)$ is the dependence of the Gaussian beam width on the distance covered:

$$w(z) = w\sqrt{1+\frac{z^2}{z_R^2}}, \tag{1.135}$$

$x_0(z)$ is the dependence of the coordinate of the centre (intensity maximum) of the Gaussian beam on the distance travelled:

$$x_0(z) = -\frac{\alpha z^2}{4}, \tag{1.136}$$

$x_1(z)$ is the dependence of the coordinate of the centre of curvature of the Gaussian beam on the distance travelled:

$$x_1(z) = -\left[x_0(z) - \frac{\alpha z_R^2}{2} \right], \tag{1.137}$$

$\zeta(z)$ if the Gouy phase:

$$\zeta(z) = -\arctan\left(\frac{z}{z_R} \right), \tag{1.138}$$

$R(z)$ is the dependence of the radius of curvature of the Gaussian beam wave front on the distance travelled:

$$R(z) = z\left[1 + \left(\frac{z_R}{z} \right)^2 \right], \tag{1.139}$$

$\Phi(z)$ is the additional phase shift:

$$\Phi(z) = \frac{k\alpha^2 z^2}{24}\left[2z - 3R(z) \right]. \tag{1.140}$$

The expressions (1.134)–(1.140) show that in a gradient medium with a linear dependence of the dielectric constant of the transverse coordinate (1.114) with the paraxial propagation of the Gaussian beam its centre moves along a parabola in proportion to z^2, and its radius coincides with the radius of the Gaussian beam propagating in a homogeneous medium with a refractive index of n_0.

Numerical simulation of the propagation of the Gaussian beams in a medium (1.114) was carried out using the FDTD finite-difference method for solving the Maxwell equations. A gradient medium with a dielectric constant was considered, which in the modelling domain changed linearly from $\varepsilon_a = 1$ (air) to $\varepsilon_g = 2.25$ (glass) (Fig. 1.17). Other simulation parameters were selected as follows: the wavelength of light in vacuum $\lambda = 633$ nm, the simulation domain: $-20\lambda \leq x \leq 20\lambda$, $0 \leq z \leq 50\lambda$, simulation time $0 \leq t \leq 100\lambda/c$ (c is the wavelength of light in vacuum), the discretization step in both coordinates is $\lambda/16$, and in time $\lambda/32$. The dielectric constant at the centre is equal to $\varepsilon(x = 0) = (\varepsilon_a + \varepsilon_g)/2 = 1.625$ (i.e. $n_0 = 1.27$). Parameter α was chosen based on the conditions $\varepsilon(x = -20\lambda) = \varepsilon_g$, $\varepsilon(x = +20\lambda) = \varepsilon_a$ and therefore $\alpha = 1/52\lambda \approx 0.03$ μm^{-1}. The Gaussian beam waist radius was $w = 2\lambda$, polarization – TE, i.e. $\mathbf{E} \equiv (0, E_y, 0)$.

Figure 1.18 *a* shows the time-averaged intensity in the *Oxz* plane. The light points mark the centres of the Gaussian beam at various distances z calculated by formula (1.135). For comparison Fig. 1.18 *b* shows the time-averaged intensity of a Gaussian beam propagating in a homogeneous medium with a refractive index $n_0 = 1.27$. From the comparison of Figs. 1.18 *a* and *b* it can be seen that at the same distances z, the radii of the Gaussian beams coincide, i.e. formula (1.135) describes the beam radius not only in a homogeneous medium, but also in a gradient medium (1.114).

Let us 'push' the radius of the waist of the Gaussian beam to infinity and add a linear phase gradient in the input plane, i.e. when $z = 0$, the complex amplitude takes the form:

$$U(x, z = 0) = \exp(ik\beta x), \tag{1.141}$$

where β is a coefficient characterizing the angle of inclination of the flat wave. Substituting (1.141) into the integral transformation

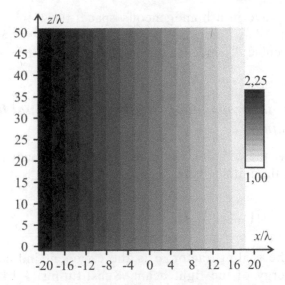

Fig. 1.17. The distribution of the dielectric constant of the medium gradient, increasing linearly from 1 (air, white, $x = 20\lambda$) and 2.25 (glass, black, $x = -20\lambda$).

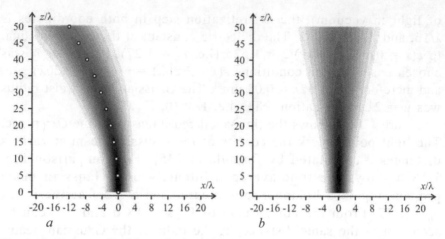

Fig. 1.18. Time-averaged intensity in the *Oxz* plane of a Gaussian beam in a medium (1.114) (*a*) and in a homogeneous medium with a refractive index $n_0 = 1.27$ (*b*).

(1.132), we obtain that in an arbitrary transverse plane at a distance z the amplitude takes the form:

$$U(x,z) = \exp\left\{ ik\left[-\frac{\alpha^2 z^3}{24} + \frac{\alpha\beta z^2}{4} - \frac{(\alpha x + \beta^2)z}{2} + \beta x \right] \right\}. \qquad (1.142)$$

Like a plane wave in a homogeneous space, the field (1.142) in a medium (1.114) has a constant intensity. An interesting feature of (1.142) is the cubic dependence of the phase on the distance travelled z.

1.4.3. Airy beam propagation in a two-dimensional linear gradient medium

Let us consider the complex amplitude of the Airy beam with limited energy [7] in the initial plane:

$$U(x,z=0) = \text{Ai}\left(\frac{x}{x_0} \right) \exp\left(a\frac{x}{x_0} \right), \qquad (1.143)$$

where x_0 is the scaling factor, a is the exponent indicator, which limits the energy of the light beam. Substituting (1.143) into the integral transform (1.133), we obtain that in an arbitrary transverse plane at a distance z the amplitude takes the form:

$$U(x,z) = \exp\left[i\left(\frac{\alpha}{2} - \frac{k^2 x_0^3 \alpha^2}{6} - \frac{1}{3k^2 x_0^3}\right)\frac{z^3}{4kx_0^3}\right] \times$$

$$\times \exp\left[\frac{a}{2x_0}\left(\frac{\alpha}{2} - \frac{1}{k^2 x_0^3}\right)z^2\right] \times$$

$$\times \exp\left[i\left(\frac{x}{x_0} - k^2 \alpha x_0^2 x + a^2\right)\frac{z}{2kx_0^2} + a\left(\frac{x}{x_0}\right)\right] \times$$

$$\times \mathrm{Ai}\left[\frac{x}{x_0} + \left(\alpha - \frac{1}{k^2 x_0^3}\right)\frac{z^2}{4x_0} + \frac{iaz}{kx_0^2}\right].$$

(1.144)

In the general case, the argument of the Airy function depends on both Cartesian coordinates x and z; however, when passing to the unbounded Airy beam ($a = 0$), all exponents in (1.144) become pure-phased and it is easy to see that the Airy beam propagates along a parabolic trajectory. When matching the parameters x_0 and α, when $\alpha = 1/(k^2 x_0^3)$, the argument of the Airy function loses its dependence on z and the amplitude (1.144) takes the form:

$$U(x,z) = \mathrm{Ai}\left(\frac{x}{x_0}\right). \tag{1.145}$$

The amplitude (1.145) corresponds to the modal solution of equation (1.115), described in [75, 76]. From (1.145) it can be seen that the Airy beams whose scale is consistent with the properties of the medium, propagate in a linear-gradient medium (1.114) in a linear fashion (Fig. 1.19). Figure 1.19 was obtained by modelling with the FDTD method, the simulation parameters are the same as in Fig. 1.18, however, the simulation domain on the x axis was extended $-40\lambda \le x \le 40\lambda$. A small bend of the beam path in Fig. 1.19 is a consequence of the boundedness of the beam in the input plane.

The following results are obtained in the section. An integral transform (1.132) is obtained, which describes the paraxial propagation of the light beam in a planar gradient medium with a linear dependence of the dielectric constant on the transverse coordinate. It is shown that the propagation of light at a distance z in such a medium is equivalent to passing through a prism whose strength is proportional to z, subsequent spreading in a homogeneous medium with a refractive index n_0, repetitive passing through the said prism, and an additional constant phase shift depending on

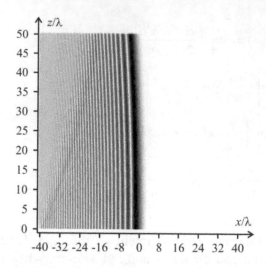

Fig. 1.19. Time-averaged intensity in the *Oxz* plane of the rectilinear Airy beam in a medium (1.114).

the distance z. It is shown that when a Gaussian beam propagates in a gradient medium (1.114), its centre shifts along a parabola, proportional to z^2, and its radius coincides with the radius of a Gaussian beam propagating in a homogeneous medium with a refractive index on the optical axis of the medium (1.114). Using the obtained integral transform (1.132) and the initial field (1.143), we obtained an explicit expression for the complex amplitude of the Airy beam in a planar medium with a linear profile (1.144), which coincides with the expression obtained in [39] in another way.

1.5. Accelerating 2D Bessel beams

After the discovery of the Airy beams with finite energy [7], interest in light beams with a curved trajectory increased. The Airy beams propagate along a parabolic trajectory and demonstrate such properties as diffraction-free and self-healing after distortion by obstacles. The Airy beams are still actively investigated: symmetric Airy beams are considered in [77], the effect of astigmatism on the type of Airy beam is studied in [78]. A three-dimensional Airy beam with an arbitrary angle between the 'two wings' was studied in [79], and a spiral focussing of Airy beams with an optical vortex was considered in [80]. In addition to the Airy beams, other nonparaxial accelerated light beams with arbitrary transverse form [81] were studied. However, the Airy beams are paraxial, their complex amplitude is a solution of the Helmholtz equation in the paraxial approximation (Schrödinger-type equations). In this regard, in recent years, attempts have been made to overcome the paraxial constraint and find solutions of the exact Helmholtz equation describing accelerating light beams [44, 82] that can rotate when propagating at greater angles than the paraxial beams. From any exact solution of the three-dimensional Helmholtz equation describing the diffraction-free light beam, one can obtain a solution of the two-dimensional Helmholtz equation. To do this, it is enough to take the propagation constant equal to zero. So, in work [82] decisions based on Mathieu and Weber beams were taken as the basis, and in [44] Bessel modes, but the angular spectrum of plane waves was truncated, which led to the formation of a light beam with a trajectory in the form of a semiring. In recent work [83], it was proposed to transform the light field from a point source into a field with a circular caustic using a mirror that is longer than a half ring. But the complete ring has not as yet been produced. In this section, it is proposed to form a two-dimensional Bessel beam in such a way as to form a light ring through which light circulates continuously.

Another drawback of the Airy beams, like many other light beams, is that the intensity distribution in the cross section has, in addition to the main maximum, auxiliary maxima (side lobes). A number of papers deal with methods for suppressing the side lobes. Thus, in [84] it was proposed to use a conical axicon to suppress these petals in the far zone of the diffraction of a plane wave on a spiral phase plate bounded by a circular diaphragm. In [85], on the basis of the Debye vector formulas, an annular amplitude mask was calculated for suppressing the side lobes of a radially polarized sharply focused Laguerre–Gaussian beam. Work [86] uses asymmetric apodization of the Fourier spectrum to enhance the main intensity maximum and suppress the side lobes in the Airy beams.

In this section we, by analogy with the work [44], consider a two-dimensional Bessel beam, but by using two sources, located opposite each other, will show the possibility of formation of a light ring which can be regarded as the accelerating light beam rotating through an angle of 360°. In addition, the solution of the two-dimensional Helmholtz equation is considered below, which, as well as the beams in [44, 82], is obtained from the solution of the three-dimensional Helmholtz equation describing the diffraction-free beam. But unlike [44, 82], instead of the standard Bessel beam, the solution for the asymmetric Bessel beam was taken as the basis [87, 88]. It is shown that this solution can be obtained from the traditional Bessel beam by a real shift along one Cartesian coordinate, and an imaginary shift along the other, as proposed in [89]. It is also shown that with an increase in the asymmetry parameter of the light beam, the complex amplitude of which is described by the solution found, its propagation distance decreases, but the intensity of the side lobes also decreases.

1.5.1. Modelling the formation of a ring 2D Bessel beam

The two-dimensional Helmholtz equation has the form:

$$\frac{\partial^2 E}{\partial r^2} + \frac{1}{r}\frac{\partial E}{\partial r} + \frac{\partial^2 E}{\partial \varphi^2} + k^2 E = 0, \tag{1.146}$$

where (r, φ) are the polar coordinates in the plane Oxz ($x = r \cos \varphi$, $z = r \sin \varphi$), $k = 2\pi/\lambda$ is the wave number for light with wavelength λ. In separable variables, this equation has the solution

$$E_n(r,\varphi) = J_n(kr)\exp(in\varphi), \tag{1.147}$$

where n is an integer (the order of the Bessel mode). The light field with a complex amplitude (1.147) can be represented as a superposition of plane waves:

$$E_n(x,z) = \frac{(-i)^n}{2\pi} \int_0^{2\pi} \exp(in\theta + ikx\cos\theta + ikz\sin\theta)\,d\theta. \tag{1.148}$$

It is easy to see that the field (1.148) consists of plane waves that have all possible directions, since the angle θ, which determines the direction of their propagation, varies from 0 to 2π. The field (1.148) can be represented as:

$$E_n(x,z) = E_n^+(x,z) + E_n^-(x,z), \tag{1.149}$$

where

$$E_n^\pm(x,z) = \frac{(\mp i)^n}{2\pi} \int_0^{\pi} \exp(in\theta \pm ikx\cos\theta \pm ikz\sin\theta)\,d\theta. \tag{1.150}$$

Field (1.150) can be represented as a series of Bessel, Anger and Weber functions:

$$E_n^{\pm}(x,z) = \frac{(\mp i)^n}{2} \sum_{m=-\infty}^{+\infty} (\pm i)^m J_m(kx) \left[\mathbf{J}_{n-m}(\mp kz) + \mathbf{E}_{n-m}(\mp kz) \right], \quad (1.151)$$

where $\mathbf{J}_v(z)$ and $\mathbf{E}_v(z)$ are the Anger and Weber functions:

$$\mathbf{J}_v(z) = \frac{1}{\pi} \int_0^\pi \cos(v\theta - z\sin\theta) d\theta,$$

$$\mathbf{E}_v(z) = \frac{1}{\pi} \int_0^\pi \sin(v\theta - z\sin\theta) d\theta. \quad (1.152)$$

The field $E_n^-(x,z)$ propagates in the negative direction of the z axis. This means that the field (1.147) is shifted by some value z_0 along the optical axis, i.e. the field of the type

$$E_n(x,z) = J_n\left(k\sqrt{x^2 + (z-z_0)^2} \right) \exp\left[in \arctan\left(\frac{z-z_0}{x} \right) \right], \quad (1.153)$$

is impossible to form if in the initial plane $z = 0$ the complex amplitude is specified as

$$E_n(x,0) = J_n\left(k\sqrt{x^2 + z_0^2} \right) \exp\left[-in \arctan\left(\frac{z_0}{x} \right) \right]. \quad (1.154)$$

In this case, instead of the light ring, only a part of it will be formed (almost a half ring). So, Fig. 1.20 shows the intensity distributions in the Oxz plane, calculated using integral (1.148), but in Fig. 1.20 a and the lower and upper limits of integration are 0 and 2π respectively, and in Fig. 1.20 b 0 and π (i.e. the value $|E_n^+(x,z)|^2$ is shown in Fig. 1.20 b). Figure 1.20 was obtained for $\lambda = 532$ nm, $n = 20$, $-2\lambda \le x, z \le 2\lambda$.

Therefore, to form a light ring, it is required that the plane waves propagate along the optical axis both in the positive direction and in the negative direction.

If the initial field is set in the plane passing through the centre of the ring, then the half ring will be formed as shown in Fig. 1.21

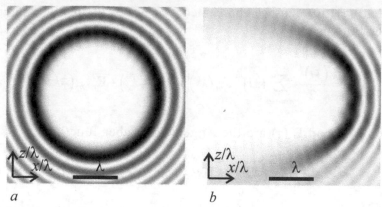

a   *b*

Fig. 1.20. The distribution of intensity in the plane Oxz, calculated using the integral (1.148) for different values of the limits of integration from 0 to 2π (a), from 0 to π (b).

a. Figure 1.21 *a* was obtained by numerical simulation using the finite-difference FDTD method, when the initial field was specified as (TE-polarization):

$$E_y\left(x, z=0\right)=J_n\left(kx\right) \tag{1.155}$$

when $n = 20$. The remaining parameters were equal to: modelling domain: $-60\lambda \leq x$, $z \leq 60\lambda$, $0 \leq z \leq 40\lambda$, the modelling time 100 λ/c, the sampling pitch in both coordinates $\lambda/32$, and in time $\lambda/(64c)$.

If the initial field is displaced from the centre of the ring and specified in the form (1.154), then the half ring shown in Fig. 1.21 *b* will be formed.

In order to form a ring instead of a half ring, the counter-propagation of two Bessel beams (1.153) should be ensured. In this case, a light ring is actually formed, however, as a result of interference on the ring, significant intensity dips are pronounced (Fig. 1.21 *c*).

To form a ring with a constant intensity, the propagation of fields (1.149) from two sources located at a distance of 10λ opposite each other was also simulated. In the $z = 0$ plane, the complex amplitude was given as

$$E\left(x, z=0\right)=E_n^+\left(x,-z_0\right)=\frac{\left(-i\right)^n}{2\pi}\times$$
$$\times\int\limits_0^\pi \exp\left(in\theta + ikx\cos\theta - ikz_0\sin\theta\right)d\theta, \tag{1.156}$$

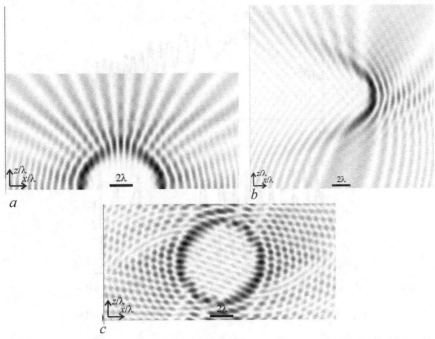

Fig. 1.21. The distribution of the time-averaged intensity of the light field in the *Oxz* plane with the initial complex amplitude (1.155) (*a*) and (1.154) (*b*), as well as in the case of the counter-propagation of two light fields (1.152) (*c*).

and in the $z = 2z_0 = 10\lambda$ plane, the complex amplitude was defined symmetrically in the form $E(x, z = 2z_0) = E(-x, z = 0)$. The amplitude distribution and that of the phase of the field (1.156) shown in Fig. 1.22.

These two linear sources generate beams propagating towards each other and form a two-dimensional Bessel beam with a centre at $x = 0$, $z_0 = 5\lambda$. The other parameters were equal to: modelling domain: $-100\lambda \leq x$, $z \leq 100\lambda$, $0 \leq z \leq 10\lambda$, the modelling time $100\lambda/c$, the sampling pitch in both coordinates $\lambda/64$, and in time $\lambda/(128c)$. Figure 1.23 shows the time-averaged intensity of the formed two-dimensional Bessel beam with a topological charge $n = 20$. The remaining small unevenness of the intensity distribution on the light ring is explained by the finite dimensions of both light sources.

Figure 1.24 shows the distribution of the power flux in the $z = 5\lambda$ plane (Fig. 1.24 *a*) and in the $x = 0$ plane (Fig. 1.24 *b*).

From Fig. 1.24 it follows that the light power flux propagates through the light ring (Fig. 1.23) counterclockwise.

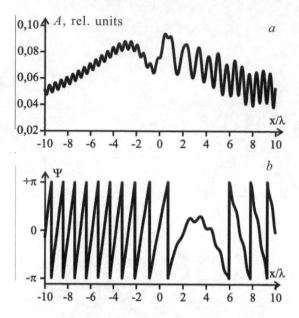

Fig. 1.22. Amplitude A and phase Ψ of the field (1.156).

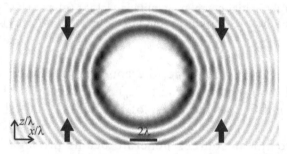

Fig. 1.23. The time-averaged intensity of a two-dimensional Bessel beam calculated by the FDTD method (the arrows indicate the directions of light propagation from sources).

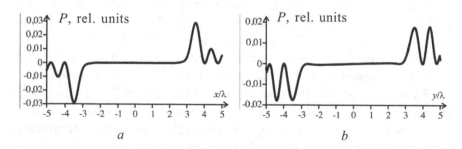

Fig. 1.24. Distribution of power flux in planes $z = 5\lambda$ (*a*) and $x = 0$ (*b*).

1.5.2. Modelling of the propagation of a two-dimensional asymmetric Bessel beam

The asymmetric nonparaxial Bessel modes, which are an exact solution of the three-dimensional scalar Helmholtz equation, were considered in [87, 88]. A similar solution of the two-dimensional Helmholtz equation can be obtained if the transverse coordinate y is replaced by the longitudinal coordinate z, and the scaling parameter of the Bessel beam is taken equal to the wave number k:

$$E_n(x,z;c) = \left[\frac{kr}{kr - 2c\exp(i\varphi)}\right]^{n/2} \times$$
$$\times J_n\left\{\sqrt{kr[kr - 2c\exp(i\varphi)]}\right\}\exp(in\varphi), \tag{1.157}$$

where (r, φ) are the polar coordinates in the Oxz plane. Turning to the Cartesian coordinates, we obtain the expression for the complex amplitude:

$$E_n(x,z;c) = \left[\frac{k(x+iz)^2}{k(x^2 + z^2) - 2c(x + iz)}\right]^{n/2} \times$$
$$\times J_n\left\{\sqrt{k^2(x^2 + z^2) - 2kc(x + iz)}\right\}. \tag{1.158}$$

Note that the beam (1.158) coincides with (1.155) with $c = 0$ and $z = 0$.

There is a method of obtaining solutions of the Helmholtz equation by the complex source offset method. This idea originated in 1967, when Kravtsov [89] showed that the field of a point source with an imaginary location in the paraxial approximation reduces to a paraxial Gaussian beam. Later this method was developed in [90–93]. In this work, we show that the solution (1.158) can also be obtained from the complex amplitude of a conventional Bessel beam, but whose axis is shifted along the x axis by a real value, and along the y axis by an imaginary one. To do this, we introduce two complex coordinates:

$$\begin{cases} u = x - c/k, \\ v = z - ic/k. \end{cases} \tag{1.159}$$

Then (1.158) is transformed as follows:

$$E_n\left(u,v\right)=\left(\frac{u+iv}{\sqrt{u^2+v^2}}\right)^n J_n\left(k\sqrt{u^2+v^2}\right). \qquad (1.160)$$

If the coordinates u and v have to be real, the expression (1.160) would describe the complex amplitude of the traditional Bessel beam in the Cartesian coordinates.

Figure 1.25 shows the distribution of the time-averaged intensity in the *Oxz* plane of the two-dimensional Bessel beam, and Fig. 1.26 is the intensity distribution of the same beam in the $z = 0$ plane.

Figures 1.25 and 1.26 were obtained by numerical finite-difference simulation FDTD-method (Finite difference in time-domain) solutions of Maxwell's equations for the following values of parameters: wavelength $\lambda = 532$ nm, the order of the Bessel function $n = 40$, simulation domain $-60\lambda \le x \le 60\lambda$, $0 \le z \le 40\lambda$, the discretization step in both Cartesian coordinates is $\lambda/32$, in time $- \lambda/(64c)$, where c

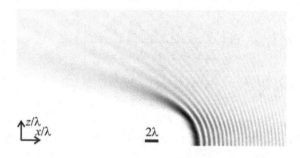

Fig. 1.25. The distribution of time-averaged intensity of the beam (1.160) in the plane *Oxz*.

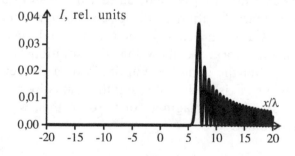

Fig. 1.26. The intensity distribution of the beam (1.161) in the $z = 0$ plane.

is the speed of light in a vacuum. The following complex amplitude (TE-polarization) was set as the initial field in the $z = 0$ plane:

$$E_y\left(x, z = 0\right) = \begin{cases} J_n\left(kx\right), & x \geq 0, \\ 0, & x < 0. \end{cases} \tag{1.161}$$

Figures 1.25 and 1.26 show that the field (1.161) has many side maxima, in addition to the main intensity maximum. The intensity of the second maximum is 57% of the intensity of the main maximum.

In order to reduce the contrast of the side lobes, instead of the field (1.161), one can use the field (1.158) with $z = 0$ and different values of the asymmetry parameter c. In the initial plane, the complex amplitude of such a field is:

$$E_n\left(x, z = 0;\ c\right) = \left(\frac{kx}{kx - 2c}\right)^{n/2} J_n\left\{\sqrt{kx\left(kx - 2c\right)}\right\}. \tag{1.162}$$

Figures 1.27 *a* and *b* shows the distribution of the time-averaged intensity in the *Oxz* plane of the two-dimensional asymmetric Bessel beam with the initial field (1.162) for the following asymmetry parameter values: $c = 1$ (Fig. 1.27 *a*) and $c = 2$ (Fig. 1.27 *b*). Values

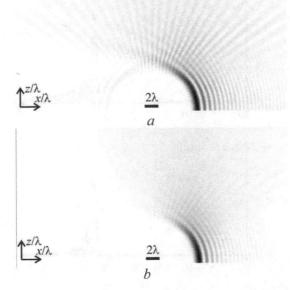

Fig. 1.27. Distributions of the beam intensity (1.158) averaged over time in the *Oxz* plane with $c = 1$ (*a*) and $c = 2$ (*b*).

of other parameters are the same as in Figs. 1.25 and 1.26. Figure 1.28 shows the intensity distribution of the same beams in the $z = 0$ plane.

The intensity of the second maximum in Fig. 1.28 *a* is 45% of the intensity of the main maximum, and in Fig. 1.28 *b* it is 35%. The intensity of the second maximum with respect to the main one was also calculated for $c = 0.5$, the obtained dependence turned out to be almost linear and is shown in Fig. 1.29. From Fig. 1.29 it follows that the value of the second maximum can be reduced from 57% to 20%, i.e. almost 3 times.

From the comparison of Figs. 1.25 and 1.27 it can be seen that the beam propagation length decreases with increasing asymmetry parameter *c*. We define this length as the distance z_1, at which the intensity is halved:

$$\max_{x}\left|E(x,z_1)\right| = \frac{1}{2}\max_{x}\left|E(x,0)\right|. \tag{1.163}$$

Figure 1.30 shows the dependence of the beam propagation length on the asymmetry parameter *c*.

Figure 1.30 shows that, although an increase in the asymmetry parameter *c* of the two-dimensional Bessel beam (1.158) decreaes

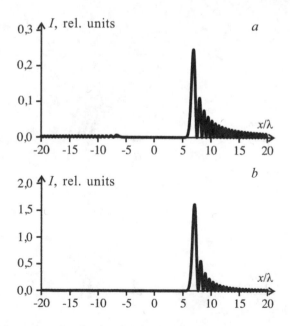

Fig. 1.28. The intensity distribution of the beam (1.157) in the $z = 0$ plane with $c = 1$ (*a*) and $c = 2$ (*b*)

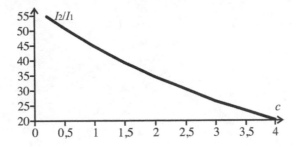

Fig. 1.29. The dependence of the intensity of the second maximum to the intensity of the main maximum on the asymmetry parameter c.

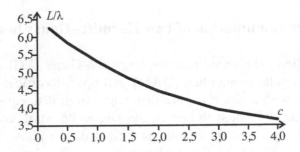

Fig. 1.30. Dependence of the length of the light beam (on the half-decay of the intensity) on the asymmetry parameter c.

the relative magnitude of the side lobes, the length of the trajectory also decreases (but more slowly).

The following results are obtained in this section. It is shown that if two two-dimensional beams the complex amplitude of which is described by a series of the Bessel, Anger and Weber functions, are directed towards each other, then a light ring will be formed through which the power flux circulates (equations (1.149)–(1.151)). It is shown analytically that the asymmetric Bessel mode is obtained by displacing the traditional Bessel mode along the Cartesian coordinates, and the displacement along one coordinate is an imaginary quantity (equation (1.160)). The solution of the two-dimensional Helmholtz equation describing a two-dimensional accelerated asymmetric Bessel beam was obtained (equation (1.157)). It has been shown that the beam propagation distance decreases with an increase in the asymmetry parameter of a two-dimensional asymmetric Bessel beam, but the side lobe size also decreases relative to the main lobe (Figs. 1.29 and 1.30).

Hermite–Gauss vortex beams

2.1. Linear combination of two Hermite–Gauss beams

In 1992, Allen [9] showed that the Laguerre–Gauss (LG) modes have an orbital angular momentum (OAM). All optical vortices or singular laser beams with a phase feature or a wave front dislocation have the same moment [94]. In such beams, the Umov–Poynting vector (power flux) rotates in a spiral around the points of the phase singularity. The first generation of a laser beam with a phase singularity was carried out in 1979 [95]. Two Hermite–Gauss (HG) modes HG_{01} and HG_{10} interfering in the resonator of a krypton ion laser, were shown to form the LG_{01} mode. In 1989, the term optical vortex was introduced in [96]. In 1990, Soskin [97] formed an optical vortex using an amplitude diffraction grating with a 'fork', and in 1992 a singular laser beam was formed using a spiral phase plate [98].

In 1991, Volostnikov, using an astigmatic mode converter, transformed the high-order mode of the HG to the LG mode, which has phase singularity [99]. The interference mode $\pi/2$ converter was considered in [100]. The orbital angular momentum of laser beams is used to rotate microparticles, in quantum telecommunications, microscopy, interferometry, metrology, etc. A modern overview of OAM for light can be found in [101]. Also, many applications of vortex laser beams with an OAM are described in the monograph [10]. The papers [102–104] are devoted to the formation of optical vortices using interferometers. In these works a matrix of optical vortices was formed using the interference of 3 plane waves,

In [99], a formula was obtained that allows one to obtain the LG mode as a finite sum of HG modes. For example, in order to get an LG mode with the topological charge 2, it is required to add at least three HG modes with certain complex coefficients. In this book, we show that it is possible to obtain a light field with any integer OAM, adding up only two HG modes with specific numbers.

It should be noted that recently a laser optical vortex with the highest OAM and the highest topological charge equal to 5050 was formed using an aluminium reflecting diffractive optical element [105]. There are various methods of forming laser beams with fractional OAM [3, 106, 107]. In [106], laser beams with a half-integer OAM were formed using a linear combination of the LG modes. In [3], Hermite–Laguerre–Gauss beams with fractional OAM were formed with the help of an astigmatic mode converter. In [107] conical diffraction using a Gaussian beam with circular polarization was used to form a Bessel beam with the OAM smaller than unity.

In this section, the OAM is calculated for a linear combination of two HG beams, the double numbers of which consist of direct and rearranged adjacent integer numbers, and which (the beams) have a phase shift between each other of $\pi/2$. Moreover, generalized HG beams are considered, which, with certain parameters, are transferred to HG modes and elegant HG beams. It is shown that the OAM module for two corresponding HG modes is an integer, and the OAM module for two elegant HG beams is always equal to one. When adding two corresponding hybrid HG beams, the OAM module will always be a fractional number. The exception is the trivial case, when two generalized HG beams with numbers (0,1) and (1,0) are added. In this case, as for the HG modes, the OAM module is equal to 1.

2.1.1. Generalized Hermite–Gauss laser beams

The Hermite–Gauss modes have long been known in optics [60]. Elegant HG beams are described by functions with a complex argument. These beams were first considered in 1973 by Siegman [108].

However, there are HG beams which are also solutions of the paraxial equation of propagation and have an explicit analytical form. For certain parameters, these beams go over into HG modes [60] and elegant HG beams [108].

Let the complex amplitude of light in the initial plane z = 0 have the form $E_{nm}(x,y) = E_1(x)E_2(y)$, where $E_1(x) = \exp[-(x/a)^2]H_n(x/b)$ and $E_2(y) = \exp[-(y/c)^2]H_m(y/d)$, where a, b, c, d are real numbers. Since the complex amplitude is the product of two functions depending on different Cartesian coordinates, the propagation of the entire 2D beam can be considered as the propagation of the 1D beam along any of the transverse coordinates. The complex amplitude of such a 1D light field at a distance z in the paraxial approximation is calculated using the Fresnel transformation and is equal to:

$$E_1(x,z) = \left(\frac{-iz_0}{z}\right)^{1/2} i^n \left(1 - \frac{iz_0}{z}\right)^{-(n+1)/2} \left[\left(\frac{a}{b}\right)^2 - 1 + \frac{iz_0}{z}\right]^{n/2} \times$$

$$\times \exp\left[-\left(\frac{x}{a(z)}\right)^2 + \frac{ikx^2}{2R(z)}\right] H_n\left(\frac{x}{b(z)}\right), \tag{2.1}$$

where

$$z_0 = \frac{ka^2}{2},$$

$$a(z) = a\left[1 + \left(\frac{z}{z_0}\right)^2\right]^{1/2},$$

$$R(z) = z\left[1 + \left(\frac{z_0}{z}\right)^2\right], \tag{2.2}$$

$$b(z) = b\left(\frac{z}{z_0}\right)\left(1 - \frac{iz_0}{z}\right)^{1/2}\left[\left(\frac{a}{b}\right)^2 - 1 + \frac{iz_0}{z}\right]^{1/2}.$$

For the complex amplitude $E_2(y,z)$ it is possible to obtain expressions similar to (2.1) and (2.2), replacing x, n, a, b with y, m, c, d, respectively.

From (2.1), (2.2) with $a/b = c/d = \sqrt{2}$ there follows a well-known expression for the HG modes:

$$E_n(x,z) = i^n \left[\frac{a}{a(z)}\right] H_n\left[\frac{\sqrt{2}x}{a(z)}\right] \times$$

$$\times \exp\left[-\frac{x^2}{a^2(z)} + \frac{ikx^2}{2R(z)} - i(n+1/2)\arctan\left(\frac{z}{z_0}\right)\right], \tag{2.3}$$

In the two-dimensional case, the one-dimensional modes (2.3) are multiplied and the two-dimensional HG mode $E_{nm}(x,y,z) = E_n(x,z)E_m(y,z)$ is obtained.

If the condition that $a/b = c/d = 1$ is satisfied, then instead of (2.1), (2.2) we get an expression for the elegant HG modes [108]:

$$E_e(x,z) = \left(q(z)\right)^{-(n+1)/2} \exp\left[-\left(\frac{x}{aq(z)}\right)^2\right] H_n\left(\frac{x}{aq(z)}\right), \tag{2.4}$$

where $q(z) = (1 + iz/z_0)^{1/2}$. Note that the generalized (2.1) and elegant (2.4) HG beams are not free-space modes, and during propagation they change the structure of the transverse intensity distribution. It is clear that the two-dimensional generalized HG beams are formed by multiplying the corresponding functions (2.1), (2.3) and (2.4). It is possible to form a hybrid HG beam, which in one coordinate will be the HG mode, and in the other Cartesian coordinate – an elegant HG beam:

$$E_h(x,y,z=0) = \exp\left[-\left(\frac{x}{a}\right)^2 - \left(\frac{y}{c}\right)^2\right] H_m\left(\frac{\sqrt{2}x}{a}\right) H_n\left(\frac{y}{c}\right). \tag{2.5}$$

The HG beams (2.1) do not have OAM. A linear combination of HG beams with real coefficients will also have an OAM equal to zero. The nonzero OAM can be formed only for a superposition of the HG beams with complex coefficients. In the following sections, the OAM will be calculated for the superposition of two generalized HG beams with a phase delay of $\pi/2$.

2.1.2. Orbital angular momentum of a linear combination of two Hermite–Gauss modes

Let the complex amplitude of light be given in the initial plane:

$$E(x,y,0) = \exp\left[-\frac{w^2}{2}\left(x^2 + y^2\right)\right] \times$$
$$\times\left[H_{2p}(cx)H_{2s+1}(cy) + i\gamma H_{2s+1}(cx)H_{2p}(cy)\right], \tag{2.6}$$

where w, c, γ are real numbers. The OAM of such a beam can be obtained by the formula [109]:

$$J_z = \iint_{\mathbb{R}^2} E^* \left(x \frac{\partial E}{\partial y} - y \frac{\partial E}{\partial x} \right) dxdy. \tag{2.7}$$

Strictly speaking, not all OAM is shown in (2.7), but only its projection, defined up to a constant, onto the optical axis, which is averaged over the transverse plane. The OAM (2.7) is preserved as the beam propagates [109], and therefore it can be calculated in any plane, for example, when $z = 0$. Substituting (2.6) into (2.7), we get:

$$J_z = 4i\gamma c \int_{-\infty}^{+\infty} x \exp\left(-w^2 x^2\right) H_{2p}(cx) H_{2s+1}(cx) dx \times$$

$$\times \left[2p \int_{-\infty}^{+\infty} \exp\left(-w^2 y^2\right) H_{2s+1}(cy) H_{2p-1}(cy) dy - \right. \tag{2.8}$$

$$\left. -(2s+1) \int_{-\infty}^{+\infty} \exp\left(-w^2 y^2\right) H_{2p}(cy) H_{2s}(cy) dy \right].$$

Since in (2.8) there are polynomials under the integrals, the integrals can be calculated and represented by finite sums:

$$J_z = \frac{4i\pi\gamma \left[(2p)!(2s+1)! \right]^2}{w^{4(p+s+1)}} \times$$

$$\times \sum_{k=0}^{\min(p,s+1)} \frac{\left[(s+1)c^2 - kw^2 \right]\left(c^2 - w^2 \right)^{p+s-2k} \left(2c^2 \right)^{2k}}{(p-k)!(s+1-k)!(2k)!} \times \tag{2.9}$$

$$\times \left[\sum_{k=0}^{\min(s,p-1)} \frac{\left(2c^2 \right)^{2k+1} \left(c^2 - w^2 \right)^{p+s-2k-1}}{(p-1-k)!(s-k)!(2k+1)!} - \sum_{k=0}^{\min(p,s)} \frac{\left(2c^2 \right)^{2k} \left(c^2 - w^2 \right)^{p+s-2k}}{(p-k)!(s-k)!(2k)!} \right].$$

Expression (2.9) is cumbersome, and it is difficult to conclude from it when OAM will be integer, when fractional and when it is equal to zero. One can only draw certain conclusions under certain conditions. For example, if $p > s$, $c = w + \delta$, $\delta \ll w$, then subject to $(p-c)(2s+1)^{-1}w > \delta$ in (2.9) $\text{Im}(J_z) > 0$. If in (2.9) $c = w$, all sums disappear and only the items with the maximum number k remain. From (2.6) it is clear that with $c = w$ two HG modes with rearranged numbers are added. It follows from the orthogonality of the HG modes that a nonzero OAM will be only for a linear combination of modes with two consecutive numbers, that is, when $p = s$. In this case, instead of (2.9) we get:

$$J_z = -\frac{2^{4p+2} i\pi\gamma \Big[(2p+1)!\Big]^2}{w^2}. \tag{2.10}$$

In order for the OAM to be independent of the power of the laser radiation, we consider the OAM that is normalized to the intensity of the OAM. The beam power (2.6) is described by the expression ($c = w$, $p = s$):

$$I = \iint_{\mathbb{R}^2} E^* E \, dx \, dy = \frac{\pi(1+\gamma^2)}{w^2} 2^{1+4p} (2p)!(2p+1)!. \tag{2.11}$$

Therefore, the normalized OAM (OAM divided by the power of the beam) for a linear combination of two HG modes with permuted neighbouring numbers is:

$$\frac{J_z}{I} = -\left(\frac{2i\gamma}{1+\gamma^2}\right)(2p+1). \tag{2.12}$$

From (2.12) it follows that with $\gamma = 1$ the OAM of the linear combination of two HG modes

$$E_{n,n+1}(x,y) = \exp\left[-\frac{w^2}{2}(x^2+y^2)\right] \times$$
$$\times\Big[H_{2p}(wx)H_{2p+1}(wy) + iH_{2p+1}(wx)H_{2p}(wy)\Big] \tag{2.13}$$

equal to the integer:

$$\frac{J_z}{I} = -i(2p+1). \tag{2.14}$$

Note that for some values of p the OAM will be equal to an integer and for values of $\gamma \neq 1$. For example, for $\gamma = 1/2$ OAM (2.12) will be an integer as $p = 2$: $J_z/I = -4i$, with $p = 7$: $J_z/I = -12i$, etc.

For a linear combination of the HG modes of the type

$$E(x,y,0) = \exp\left[-\frac{w^2}{2}(x^2+y^2)\right] \times$$
$$\times\Big[H_n(wx)H_{n+1}(wy) + i\gamma H_{n+1}(wx)H_n(wy)\Big] \tag{2.15}$$

similarly to (2.14), one can obtain the normalized OAM for any integer n :

$$\frac{J_z}{I} = -\frac{2i\gamma(n+1)}{1+\gamma^2}. \tag{2.16}$$

Note that due to the fact that the sum of the numbers of both modes in (2.15) is the same, the linear combination (2.15) will also be a mode (the Gouy phases are the same for both modes) and during propagation will retain their appearance, changing only scale. An interesting result was obtained: mode (2.15) has the OAM, i.e. the Poynting vector locally describes a spiral in the space along the optical axis, but the beam does not rotate and extends without changing its structure.

In practice, the beam (2.6) or (2.15) with $c = w$ can be formed using a Mach–Zehnder interferometer. The HG mode $E_{nm}(x,y)$, formed at the output from the laser, is divided by a 50% mirror into two identical beams, which fall into the different arms of the interferometer. In one of the interferometer arms, using a Dove's prism, the HG mode rotates 90° $E_{mn}(x,y)$. At the exit of the interferometer, both modes are combined in one beam with a relative phase delay of $\pi/2$.

2.1.3. Orbital angular momentum of a linear combination of two elegant Hermite–Gauss beams

Here we calculate the OAM for the superposition of two elegant HG beams. To do this, put (1.168) $c = w/\sqrt{2}$ and $p = s$ and we get:

$$E_e(x,y) = \exp\left[-\frac{w^2}{2}\left(x^2 + y^2\right)\right] \times$$
$$\times \left[H_{2p}\left(\frac{wx}{\sqrt{2}}\right) H_{2p+1}\left(\frac{wy}{\sqrt{2}}\right) + i\gamma H_{2p+1}\left(\frac{wx}{\sqrt{2}}\right) H_{2p}\left(\frac{wy}{\sqrt{2}}\right) \right]. \tag{2.17}$$

Then the rated OAM, analogous to (2.12), will have the form:

$$\frac{J_z}{I} = -\left(\frac{2i\gamma}{1+\gamma^2}\right). \tag{2.18}$$

From (2.18) it can be seen that a linear combination of two elegant HG beams (2.17) will always have an OAM equal to 1 in modulus (with $\gamma = 1$) for all possible values of numbers p.

This is a very unexpected result. It turns out that the OAM (2.14) for two HG modes is determined by the maximum number of the mode participating in the linear combination. And therefore, the higher the HG mode number, the higher the OAM of the laser beam (2.13). For the elegant beams (2.17) from (2.18) it follows that the OAM is determined by the difference of two numbers of the beam participating in a linear combination. And since the difference of the numbers of the beams in (2.17) is equal to one, then the OAM is equal to 1 in modulus (for $\gamma = 1$).

For a linear combination of elegant modes with other numbers (let $k = 2l + 1$ – odd)

$$E_e(x,y) = \exp\left[-\frac{w^2}{2}(x^2 + y^2)\right] \times$$

$$\times\left[H_{2p}\left(\frac{wx}{\sqrt{2}}\right)H_{2p+k}\left(\frac{wy}{\sqrt{2}}\right) + i\gamma H_{2p+k}\left(\frac{wx}{\sqrt{2}}\right)H_{2p}\left(\frac{wy}{\sqrt{2}}\right)\right], \quad (2.19)$$

we will receive the OAM in the form

$$\frac{J_z}{I} = -\left(\frac{i\gamma}{1+\gamma^2}\right)\frac{k(4p+k)\Gamma^2(2p+k/2)}{\Gamma(2p+1/2)\Gamma(2p+k+1/2)}, \quad (2.20)$$

where $\Gamma(x)$ is a gamma function. Expression (2.20) coincides with (2.18) with $k = 1$. With the next number $k = 3$ from (2.20) it follows:

$$\frac{J_z}{I} = -\left(\frac{2i\gamma}{1+\gamma^2}\right)\frac{3(4p+1)}{(4p+5)}. \quad (2.21)$$

At $\gamma = 1$ and for large p the OAM modulus (2.21) is close to 3, that is, close to the difference between the numbers of the elegant HG beam (2.19).

2.1.4. Orbital angular momentum of a linear combination of two Hermite–Gauss hybrid beams

By hybrid HG beam we here mean a beam that is an HG mode along one transverse Cartesian axis and is an elegant HG beam along the other transverse Cartesian axis.

A sum of two hybrid HG beams with the HG mode having a larger number than the elegant HG beam has the following form

$$E_{h1}(x,y) = \exp\left[-\frac{w^2}{2}(x^2+y^2)\right] \times$$

$$\times\left[H_{2p}\left(\frac{wx}{\sqrt{2}}\right)H_{2p+1}(wy) + i\gamma H_{2p+1}(wx)H_{2p}\left(\frac{wy}{\sqrt{2}}\right)\right]. \tag{2.22}$$

Then the normalized OAM for the beam (2.22) will have the form:

$$\frac{J_z}{I} = \left(\frac{2i\gamma}{1+\gamma^2}\right)\frac{(2p+1)!}{(4p-1)!!}. \tag{2.23}$$

Since the numerator in (2.23) contains $(2p+1)!$ and since the denominator contains $(4p - 1)!!$, the numerator includes even and odd factors, whereas the denominator includes only odd factors. Therefore, the OAM (2.23) with $\gamma = 1$ will never be an integer, except in the trivial case: $\gamma = 1$ and $p = 0$. For example, with $p = 2$ and $\gamma = 1$ from (2.23) we get $J_z/I = i8/7$. This OAM in modulus is a little more than one. We note also that the OAM (2.23) has the opposite sign, in relation to all the previous calculated OAMs.

In another case, when in the sum of two hybrid HG beams, the HG mode has a smaller number than the elegant HG beam:

$$E_{h2}(x,y) = \exp\left[-\frac{w^2}{2}(x^2+y^2)\right] \times$$

$$\times\left[H_{2p}(wx)H_{2p+1}\left(\frac{wy}{\sqrt{2}}\right) + i\gamma H_{2p+1}\left(\frac{wx}{\sqrt{2}}\right)H_{2p}(wy)\right], \tag{2.24}$$

instead of (2.23) we get another expression for the OAM:

$$\frac{J_z}{I} = -\left(\frac{2i\gamma}{1+\gamma^2}\right)\frac{(2p+1)!(2p+1)(p^2-1)}{(4p+1)!!}. \tag{2.25}$$

From (2.25) it also follows that the OAM for the beam (2.24) will always be a fractional number, except in the trivial case: $\gamma = 1$ and $p = 0$. For example, with $p = 2$ and $\gamma = 1$ from (2.25) we get $J_z/I = -i40/21$. This OAM in modulus is a little less than two. We note that in all the cases considered, a change in the sign of the parameter γ leads to a change in the sign of the OAM. Note also that the common factor in (2.12), (2.18), (2.21), (2.23) and (2.25) $2\gamma/(1 + \gamma^2)$ is always less than or equal to one.

2.1.5. Numerical simulation

Forsimulation, we consider a linear combination of two generalized HG beams [61]. Let in the initial plane ($z = 0$) the light field has a complex amplitude

$$E(x,y,0) = \exp\left(-\frac{x^2+y^2}{w^2}\right) \times$$

$$\times\left[H_{2p}(bx)H_{2s+1}(cy) + i\gamma H_{2s+1}(cx)H_{2p}(by)\right]. \tag{2.26}$$

The intensity of the light beam at a wave length λ and parameter values $p = s = 2$ (i.e. Hermite polynomials of the 4th and 5th degree are used) in the initial plane (in the region of $-10 \leq x \leq 10\lambda$, $-10\lambda \leq x \leq 10\lambda$) for some values of the scaling factors b and c is shown in Figs. 2.1–2.4.

Figures 2.1 b–2.4 b show the intensity of the beams (2.21) coherently combined with an inclined plane wave:

$$I(x,y,z=0) = \left|E(x,y,0) + C\exp(i\alpha x)\right|^2, \tag{2.27}$$

where the amplitude C and the spatial frequency α were selected for better visibility of the pictures in Figs. 2.1–2.4. Figures 2.1 b–2.4 b show characteristic 'forks' among interference fringes; at the branch point of these 'forks' there are isolated intensity zeros and phase singularities. The OAM modules for the beams are equal: 5 (Fig. 2.1 a), 1 (Fig. 2.2 a), 0.95 (Fig. 2.3 a) and 0.92 (Fig. 2.4 a).

The following results were obtained in the section [110]. An expression for the complex amplitude of the generalized Hermite-Gauss paraxial beams (2.1) is obtained. With certain parameters, these beams go over into the known HG modes and elegant HG beams. The orbital angular momentum (OAM) of a linear combination of two generalized HG beams with double numbers (in direct and inverse order) from two adjacent integers and with a phase delay of $\pi/2$ is calculated. It is shown that the OAM modulus for the sum of such two HG modes is an integer proportional to the maximum HG mode number, for the sum of two elegant HG beams is always one, for the sum of two hybrid HG beams it is a fractional number.

2.2. Hermite–Gauss vortex modes

The Hermite-Gauss (HG) modes have been known in optics since 1966 [66]. Elegant HG beams are described by functions with a

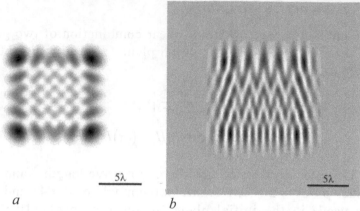

a b

Fig. 2.1. Beam intensity (2.26) without carrier frequency (a) and with carrier frequency (b) for parameter values: $w = 2\lambda$, $b = c = \sqrt{2}/w$ (HG modes).

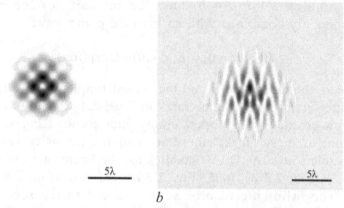

a b

Fig. 2.2. Beam intensity (2.26) without carrier frequency (a) and with carrier frequency (b) for the parameter values: $w = 2\lambda$, $b = c = 1/w$ (two elegant HG beams).

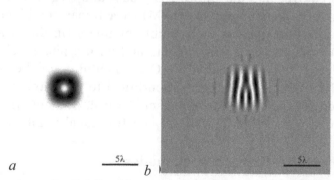

a b

Fig. 2.3. Beam intensity (2.26) without carrier frequency (a) and with carrier frequency (b) for parameter values: $w = 2\lambda$, $b = 1/(7\lambda)$, $c = 1/(3\lambda)$ (two generalized HG beams with different width along the axes)

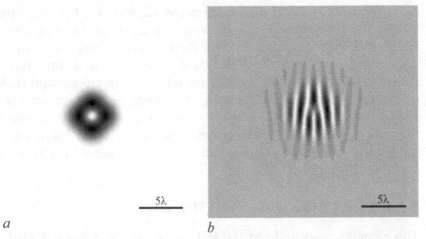

Fig. 2.4. Beam intensity (2.26) without carrier frequency (*a*) and with carrier frequency (*b*) for parameter values: $w = 2\lambda$, $b = 1/(5\lambda)$, $c = 1/(5\lambda)$ (two generalized HG beams with the same width along the axes).

complex argument. These beams were first considered in 1973 by A.E. Siegman [108]. In [111,61], generalized HG beams are considered, which are also a solution of the paraxial propagation equation and have an explicit analytic form. For certain parameters, these beams go over into HG modes [6 6] and elegant HG beams [108]. In [99], using an astigmatic mode converter a high-order HG mode was transformed into the Laguerre–Gauss (LG) mode, which has a phase singularity. The interference mode $\pi/2$ converter was considered in [100]. In [99], a formula was also obtained that allows one to obtain the LG mode as a finite sum of HG modes. For example, in order to get an LG mode with the topological charge 2, it is required to add at least three HG modes with certain complex coefficients. In [61, 110], it was shown that a light field with any integer OAM can be obtained by adding only two HG modes with certain numbers. In [3], an astigmatic mode converter was used to produce Hermite–Laguerre–Gauss beams with fractional OAM.

In this section, the Hermite–Gauss vortex modes (VHG modes) are considered. These beams are a superposition of the $(n + 1)$ th mode of the HG. The complex amplitude of VHG modes is proportional to the Hermite polynomial of degree n, the argument of which depends on the real parameter a. When $|a| < 1$ on the horizontal axis in the beam cross section there are n isolated zeros that generate optical vortices with a topological charge $+1$ ($a < 0$) or -1 ($a > 0$). When $|a| > 1$ for the VHG modes, similar isolated zeros lie on the vertical axis. When $|a| = 1$, all n isolated zeros are collected on the optical axis in the cent of the beam and generate an nth-order optical vortex and the VHG mode coincides with the Laguerre–Gauss mode of order $(0, n)$, and for $a = 0$ the VHG mode coincides with the Hermite-Gauss mode of order $(0, n)$. The orbital angular momentum (OAM) of the VHG modes was calculated; it depends on the parameter a and varies from 0 (for $a = 0$ and $a \rightarrow \infty$) to n ($a = 1$). It is shown that two modes with different numbers n and m are orthogonal, and two modes with the same number, but different values of the parameter a are not orthogonal.

2.2.1. Complex amplitude of VHG-mode

The complex amplitude of the HG mode has the form [61,110]:

$$
E_{nm}(x,y,z) = i^{n+m} \left[\frac{w}{w(z)}\right]^2 H_n\left[\frac{\sqrt{2}x}{w(z)}\right] H_m\left[\frac{\sqrt{2}y}{w(z)}\right] \times
$$
$$
\times \exp\left[-\frac{x^2+y^2}{w^2(z)} + \frac{ik(x^2+y^2)}{2R(z)}\right] \exp\left[-i(n+m+1)\operatorname{arctg}\left(\frac{z}{z_0}\right)\right], \tag{2.28}
$$

where

$$
z_0 = \frac{kw^2}{2},
$$

$$
w(z) = w\left[1+\left(\frac{z}{z_0}\right)^2\right]^{1/2}, \tag{2.29}
$$

$$
R(z) = z\left[1+\left(\frac{z_0}{z}\right)^2\right],
$$

w is the radius of the waist of the Gaussian beam, $R(z)$ is the radius of curvature of the wave front of the Gaussian beam, z_0 is the Rayleigh length, k is the wave number of light, $H_n(x)$ is the Hermite polynomial. The HG beams (2.28) do not have OAM. A linear combination of the HG beams with real coefficients will also have an OAM equal to zero. It can differ from zero only for a linear combination of the HG beams with complex coefficients [61,99]. Consider a linear combination of the HG modes (2.28) at $z = 0$ with certain coefficients:

$$U_n(x,y,z) = i^n \exp\left(-\frac{x^2+y^2}{2}\right)\sum_{k=0}^{n}\frac{n!(ia)^k}{k!(n-k)!}H_k(x)H_{n-k}(y),\qquad (2.30)$$

where a is a valid parameter. Dimensionless coordinates were introduced in (2.30) $\sqrt{2}x/w \to x$, $\sqrt{2}y/w \to y$.

Using the reference expression from [65]

$$\sum_{k=0}^{n}\frac{n!t^k}{k!(n-k)!}H_k(x)H_{n-k}(y) = \left(1+t^2\right)^{n/2}H_n\left(\frac{tx+y}{\sqrt{1+t^2}}\right),\qquad (2.31)$$

instead of (2.30), we obtain the expression for the complex amplitude of the VHG mode:

$$U_n(x,y,z) = i^n \exp\left(-\frac{x^2+y^2}{2}\right)\left(1-a^2\right)^{n/2}H_n\left(\frac{iax+y}{\sqrt{1-a^2}}\right).\qquad (2.32)$$

Since the sum of HG-mode numbers in the sum (2.30) is constant $k+(n-k)=n=$ const, then all HG modes in a linear combination (2.30) have the same phase velocity (identical Gouy ($n+m+1$) phases(z/z_0)), and therefore the entire beam (2.30) is also a paraxial mode and propagates without changing the transverse intensity structure (up to scale and rotation). But if the original HG modes (2.28), which which form the beam (2.32), have zero OAM, the beam (2.32) has OAM different from zero (if $n \ne 0$).

From (2.32) it follows that with $a = 0$ the VHG mode coincides with the usual HG mode with numbers $(0, n)$, and as $a \to \infty$, the VHG mode coincides with the ordinary HG mode with numbers $(n, 0)$. With $n = 0$ the VHG mode (2.32) coincides with the usual Gaussian beam. We find the amplitude limit (2. 32) at $a \to 1$. The term of the Hermite polynomial $H_n(x)$ has the maximum degree equal to $(2x)^n$, therefore, for $a = 1$ from (2.32) we get:

$$U_n(x,y,z) = i^n 2^n \exp\left(-\frac{x^2+y^2}{2}\right)(ix+y)^n =$$

$$= (-2)^n \exp\left(-\frac{r^2}{2}\right)r^n \exp(-in\varphi).\qquad (2.33)$$

where $r = (x^2 + y^2)^{1/2}$, $\varphi = \arctan(y/x)$.

From (2.33) it follows that the VHG mode with $a = 1$ coincides with the usual Laguerre–Gauss mode with the number $(0, n)$.

2.2.2. Orbital angular momentum of the VHG mode

We obtain the expression for the OAM of the light field (2.32). Find the OAM of such a beam using the formula [109]:

$$J_z = \operatorname{Im} \iint_{\mathbb{R}^2} E^* \left(x \frac{\partial E}{\partial y} - y \frac{\partial E}{\partial x} \right) dx\, dy. \tag{2.34}$$

Strictly speaking, not all OAM is shown in (2.34), but only its projection, determined up to a constant, onto the optical axis, which is averaged over the transverse plane. The OAM (2.34) is preserved as the beam propagates [109], and therefore it can be calculated in any plane, for example, with $z = 0$. Substituting (2.30) into (2.34), we get:

$$J_z = -\pi 2^n n! \left[2na\left(1 + a^2\right)^{n-1} \right]. \tag{2.35}$$

In deriving (2.35), we used the orthogonality property of the Hermite–Gauss functions [65]:

$$\int_{-\infty}^{\infty} \exp\left(-x^2\right) H_n(x) H_m(x)\, dx = \sqrt{\pi}\, 2^n n! \delta_{nm}, \tag{2.36}$$

where δ_{nm} is the Kronecker symbol. In order for the OAM to be independent of the power of laser radiation, we consider the OAM that is normalized to the intensity. The beam power (2.30) is described by the expression:

$$I = \iint_{\mathbb{R}^2} E^* E\, dx\, dy = \pi 2^n n! \left(1 + a^2\right)^n. \tag{2.37}$$

Therefore, the normalized OAM (OAM divided by the beam power) for the VHG mode is equal to:

$$\frac{J_z}{I} = -\frac{2an}{1 + a^2}. \tag{2.38}$$

It is interesting that the OAM (2.38) coincides with the OAM of beams consisting of a linear combination of only two HG modes, which were considered by the authors earlier in [61]. From (2.38) it follows that when $a = 1$, the OAM of the VHG-modes is equal to an integer

$$\frac{J_z}{I} = -n. \tag{2.39}$$

From (2.39) it follows that, since with $a = 1$ the VHG mode coincides with the usual Laguerre–Gauss mode with the number $(0, n)$ (2.33), then the OAM will be equal in magnitude to the topological charge of the optical vortex $\exp(in\varphi)$.

Note that for some values of n the OAM will be equal to an integer also for the values $a \neq 1$. For example, for $a = 1/2$, the OAM (2.39) will be an integer for $n = 5$: $J_z/I = -4$, for $n = 15$: $J_z/I = -12$, etc. From (2.38) it follows that for $a = 0$ and $a \rightarrow \infty$ the OAM is zero.

2.2.3. Numerical simulation

Equation (2.32) can be rewritten in the dimensional variables:

$$E_n(x, y, z = 0) = i^n \exp\left(-\frac{x^2 + y^2}{w^2}\right)\left(1 - a^2\right)^{n/2} H_n\left[\frac{\sqrt{2}\left(iax + y\right)}{w\sqrt{1 - a^2}}\right]. \tag{2.40}$$

Formula (2.40) was used to calculate the intensity and phase in the initial plane ($z = 0$) for the VHG modes with different values of the parameter a. Figure 2.5 shows the results of the calculation. The following parameters were used: the wavelength of monochromatic light $\lambda = 532$ nm, the topological charge (or Hermite polynomial number) $n = 3$, the radius of the beam waist of a Gaussian beam with an initial plane $w = 10\lambda$, size of the calculation domain $-50\lambda \leq x, y \leq +50\lambda$.

Figure 2.5 shows the intensity (*a, b, c*) and phase (*d, e, f*) of the VHG mode in the initial plane ($z = 0$) for different values of parameter a : $a = 0.5$ (*a, d*); $a = 1.5$ (*b, e*); $a = 1$ (*c, f*). Figure 2.6 *a* shows the intensity cross section from Fig. 2.5 *a* at $x = 0$. This figure shows the presence of three intensity zeros. Figure 2.6 *b* shows the intensity cross section at $a = 1.05$ and $y = 0$, while Fig. 2.6 *c* – an enlarged fragment of the cross section in the centre of

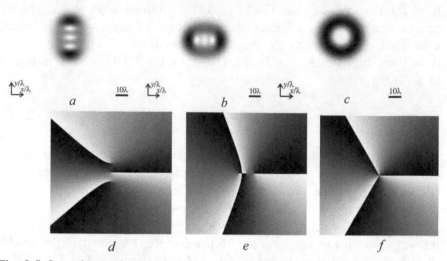

Fig. 2.5. Intensity (*a*, *b*, *c*) and phase (*d*, *e*, *f*) of VHG modes in the initial plane ($z = 0$) for different values of parameters a : $a = 0.5$ (*a*, *d*); $a = 1.5$ (*b*, e); $a = 1.001$ (*c*, *f*).

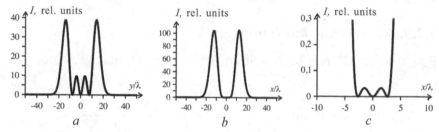

Fig. 2.6. The cross section of the intensity of the VHG mode with $a = 0.5$ and $x = 0$ (*a*), with $a = 1.05$ and $y = 0$: the whole (*b*) and enlarged fragment (*c*).

the picture (Fig. 2.6 *b*) under the same parameters. Figure 2.6 *c* also shows the presence of three zeros, not visible in Fig. 2.6 *b*. The OAM calculated by the formula (2.38) for the shown modes was equal to: $J_z/I = -2.4$ (Fig. 2.5 *a*, *d*); $J_z/I = -2.77$ (Fig. 2.5 *b*, e); $J_z/I = -3$ (Fig. 2.5 *c*, *f*). The dependence of the OAM on the parameter *a* is shown in Fig. 2.7.

Figure 2.7 shows that for a given *n* the OAM has the maximum value when $a = 1$.

2.2.4. Isolated zeros of VHG-mode intensity

We find from (2.40) the coordinates of isolated zeros of the VHG-mode intensity in the initial plane. We equate the argument of the Hermite polynomial to the root value of this polynomial γ_{nk}, that is

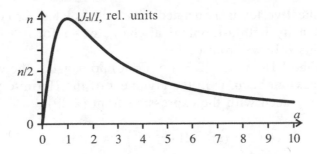

Fig. 2.7. Dependence of the OAM on parameter a.

$H_n(\gamma_{nk}) = 0$. Then we obtain the coordinates of the isolated intensity zeros:

$$\begin{cases} x_{nk} = 0, \\ y_{nk} = \gamma_{nk}\left(\dfrac{w}{\sqrt{2}}\right)\sqrt{1-a^2}, \quad |a| < 1, \end{cases} \tag{2.41}$$

and

$$\begin{cases} x_{nk} = \gamma_{nk}\left(\dfrac{w}{a\sqrt{2}}\right)\sqrt{a^2-1}, \quad |a| > 1, \\ y_{nk} = 0. \end{cases} \tag{2.42}$$

From (2.41) it is clear that for $|a| < 1$ the nth order VHG mode (2.40) has n isolated intensity zeros that lie on the y axis and have coordinates proportional to the roots of the Hermite polynomial γ_{nk}, but reduced $(1 - a^2)^{-1/2}$ times. Moreover, when $a = 0$, when the VHG mode coincides with the ordinary HG mode with the number $(0, n)$, the intensity zeros have coordinates equal to the roots γ_{nk}, and are no longer isolated intensity zeros, but belong to zero-intensity lines parallel to the x axis. With the growth of the parameter a from 0 to 1, isolated zeros (2.41) shift along the y axis to the centre $(x = y = 0)$, and for $a = 1$ all zeros 'merge' into one isolated intensity zero (n times degenerate) in the origin. From (2.42) it follows that a similar dynamics, only along the x axis, is observed for isolated zeros of the VHG mode at $|a| > 1$. As a grows, the intensity zeros 'move away' from the centre of the coordinates and, in the limit at $a \to \infty$, the coordinates of all zeros coincide with the roots γ_{nk}, and the intensity zeros themselves belong to the zero lines parallel to the y axis.

An optical vortex is associated with each isolated zero of the VHG mode intensity with a topological charge of -1 for $a > 0$ or $+1$ for $a < 0$. This follows from (2.33).

When the VHG mode (2.31) or (2.40) propagates, the waist radius of the Gaussian beam w in the argument of the Hermite polynomial must be replaced with the expression from (2.29):

$$w(z) = w\left(1 + z^2 / z_0^2\right)^{1/2}. \tag{2.43}$$

Substituting (2.43), for example, into (2.41), we obtain the coordinates of the isolated zeros:

$$\begin{cases} x_{nk} = 0, \\ y_{nk} = \gamma_{nk}\left(\dfrac{w\sqrt{1+z^2/z_0^2}}{\sqrt{2}}\right)\sqrt{1-a^2}, \ |a| < 1. \end{cases} \tag{2.44}$$

From (2.44), it follows that the isolated intensity zeros (and the intensity picture itself) of the VHG mode change with propagation only on the scale: the intensity zeros, remaining on the y axis, move away from the origin with increasing distance z.

The remarkable property of the isolated intensity zeros follows from (2.41) and (2.42): the minimum distance between them is not limited by the diffraction limit and the wavelength, but can be arbitrary. At the value of the parameter a, which differs from unity by an infinitely small value $\delta \ll 1$, the distance between the extreme isolated intensity zeros for the VHG mode (2.40) will be less than $\sqrt{2n(n-1)}w\delta$. Of course, in propagating in space with increasing z the distance between the zeros will grow: $\sqrt{2n(n-1)}w\delta\sqrt{1+z^2/z_0^2}$.

2.2.5. VHG-mode orthogonality

The dot product of the complex amplitudes of two VHG modes with different parameters n, a and m, b is as follows. It can be shown that equality holds:

$$\int_{-\infty}^{\infty} E_n\left(x,y,z=0;a\right)E_m^*\left(x,y,z=0;b\right)dx\,dy = \pi 2^n\left(n!\right)^2\left[\frac{1-(ab)^{n+1}}{1-(ab)}\right]\delta_{nm}, \tag{2.45}$$

where δ_{nm} is the Kronecker symbol. From (2.45) it follows that the VHG modes are orthogonal by the number n, and are not orthogonal by the parameter a. When $ab \rightarrow 1$, the uncertainty in (2.45) is revealed. It is interesting to note that two VHG modes with different numbers n and m will be orthogonal and will have the same OAM (but not the maximum) provided that the parameters a and b satisfy the condition:

$$\frac{n}{m} = \left(\frac{b}{a}\right)\left(\frac{1+a^2}{1+b^2}\right). \tag{2.46}$$

Physically, this is explained by the fact that the laser beam OAM depends not only on the number of isolated zeros, for example, with a topological charge +1, but also on what coordinates they have [61]. The further the intensity zeros are from the centre of the Gaussian beam, the less their contribution to the OAM. The maximum contribution to the OAM intensity zeros is given when they all 'gather' into one degenerate zero in the centre of the Gaussian beam.

2.2.6. Experiment

VHG beams were produced using the optical circuit shown in Fig. 2.8. A spatial light modulator SLM PLUTO-VIS (resolution 1920 × 1080 pixels, pixel size 8 μm) was used to display the images of the phases. The output laser beam of a solid-state laser ($\lambda = 532$ nm) was attenuated using neutral density filters F. A system made from MO micro-lens (40×, $NA = 0.6$), lens L_1 ($f = 350$ mm) and pinhole PH (hole size 40 μm) was used to obtain a uniform Gaussian intensity profile of a laser beam illuminating SLM. In addition, it allowed the beam to be expanded in order for it to completely cover the modulator diplay. Aperture D_1 allowed changing the radius of the beam incident on the display of the light modulator. The beam reflected from the modulator was directed to the lens L_2 ($f_3 = 350$ mm) by means of a beam splitter BS and a rectangular prism RP. This lens in combination with the diaphragm D_2 was used for high-frequency optical filtration. Then, using an L_3 lens ($f_2 = 150$ mm), an image was built on the matrix of a CMOS-camera MDCE-5A (1/2", resolution 1280 × 1024 pixels). The distance between the aperture D_2 and the lens L_3 was greater than the focal length of the lens, equal to 150 mm, this allowed to obtain a descending light beam, while the focus of the system was obtained at a distance of about 550 mm from the plane $z = 0$, the conjugate plane of the display of

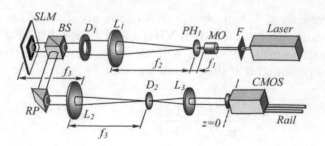

Fig. 2.8. Optical scheme for forming light VHG beams: Laser – solid-state laser (λ = 532 nm), F – neutral density filter, MO – micro-lens (40×, NA = 0.6), PH – pinhole (40 microns), L_1, L_2 – lenses with focal length $f_2 = f_3$ = 350 mm, L_3 – lens with focal length f_4 = 150 mm, BS – beam divider, SLM – spatial light modulator PLUTO_VIS, RP – rectangular prism, D – diaphragm, CMOS – CMOS camera MDCE-5 A (1280 × 1024), Rail – optical rails.

the modulator. For separation in space of the zero and first orders of diffraction, the addition of the initial phase function with a linear phase mask was used.

The image of the phase used in the experiment had a size of 1024 × 1024 pixels, so the size of the phase element displayed on the modulator display was approximately 8.2 × 8.2 mm. The diameter of the illuminating beam during the experiments was about 3 mm. These experimental parameters allowed to form a VHG beam, whose image at various distances from the SLM is shown in Fig. 2.9. The image data was generated using the phase distribution shown in Fig. 2.9 *a* (linear phase is not shown). From these images it is evident that the generated laser beam is similar to the calculated beam (Fig. 2.5 *a*) and retain their structure during propagation in the space up to scale.

Figure 2.10 shows the coded phase (n = 10), which takes into account the amplitude of the VHG beam at z = 0. Therefore, the intensity patterns recorded at different distances from the modulator (Fig. 2.10 *c, d, e*), first, also retain their appearance during propagation, and, secondly, more accurately reproduce the intensity distribution of an ideal VHG beam (Fig. 2.10 *b*).

So, this section shows that the two-parameter (n, *a*) family of VHG-modes (2.40) can be generalized to the three-parameter family (n, *a*, *b*) using the addition formula of Hermite polynomials [112, 8.958], [113, p. 254]:

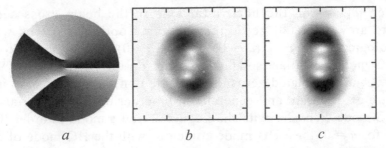

Fig. 2.9. Uncoded phase for the formation of a VHG beam ($n = 3$) (a) and experimentally formed intensity distributions (negatives) at various distances: b) 400 mm; c) 550 mm. The grid step on the images is 0.3 mm, the topological charge of the optical vortex $n = 3$.

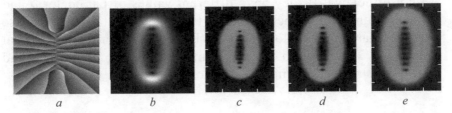

Fig. 2.10. The coded phase for the formation of the VHG-beam ($n = 10$) (a) and the calculated amplitude (b), and the experimentally formed intensity distributions (negatives) at different distances: c) 100 mm; d) 150 mm; e) 200 mm. The grid spacing on images is 0.5 mm.

$$\sum_{k=0}^{n} \frac{(ia)^k (b)^{n-k}}{k!(n-k)!} H_k(x) H_{n-k}(y) = (n!)^{-1} (b^2 - a^2)^{n/2} H_n\left(\frac{iax + by}{\sqrt{b^2 - a^2}}\right). \quad (2.47)$$

Taking into account (2.47), instead of (2.40) we will have:

$$E_{nab}(x, y, z = 0) = i^n \exp\left(-\frac{x^2 + y^2}{w^2}\right)(b^2 - a^2)^{n/2} H_n\left[\frac{\sqrt{2}(iax + by)}{w\sqrt{b^2 - a^2}}\right], \quad (2.48)$$

where a, b are real numbers. Equation (2.48) describes the complex amplitude of the generalized VHG beams, the OAM of which reaches a maximum at $a = b$. Note that replacing $a' = a/b$ reduces beams (2.48) to beams (2.40).

The following results are obtained in this section. The Hermite–Gauss vortex modes (VHG modes) are considered, whose complex amplitude is proportional to the Hermite polynomial of degree n, whose argument depends on the real parameter a.

When $|a| < 1$ on the horizontal axis in the beam cross section there are n isolated zeros that generate optical vortices with a topological charge $+1$ ($a < 0$) or -1 ($a > 0$). When $|a| > 1$ for the VHG modes, similar isolated zeros lie on the vertical axis. When $|a| = 1$, all n isolated zeros are collected on the optical axis in the centre of the beam and generate an nth-order optical vortex and the VHG mode coincides with the Laguerre–Gauss mode of order $(0, n)$, and for $a = 0$, the VHG mode coincides with the HG mode of order $(0, n)$. The orbital angular momentum (OAM) of the VHG modes was calculated, which depends on the parameter a and varies from 0 (for $a = 0$ and $a \rightarrow \infty$) to n ($a = 1$). It is shown that two VHG modes are orthogonal if they have different values of number n, and are not orthogonal if they have different values of parameter a and the same number n. Experimentally, the liquid-crystal spatial phase light modulator was used to formed VHG-beams with the topological charge $n = 3, 10$, which is consistent with the calculated one and retains its structure during propagation.

Asymmetric vortex laser beams

3.1. Asymmetric Bessel diffraction-free modes

It is known [114] that the Helmholtz equation

$$\left(\nabla^2 + k^2\right) E(x, y, z) = 0, \tag{3.1}$$

where $k = 2\pi/\lambda$ is the wave number of monochromatic light with a wavelength λ, in a cylindrical coordinate system (r, φ, z) has a solution in the form of a Bessel mode

$$E_n(r, \phi, z) = \exp\left(ikz\cos\theta_0 + in\phi\right) J_n(kr\sin\theta_0), \tag{3.2}$$

where θ_0 is the angle of the conical wave forming the Bessel beam, $J_n(x)$ is the n-th order Bessel function of the first kind. The angle θ_0 determines the amplitude of the spectrum of plane waves of the Bessel mode (3.2) on the unit sphere $F_n(\theta, \phi) = (-i)^n \exp(in\phi)\delta(\theta - \theta_0)$, where $\delta(x)$ is the Dirac delta function. The Bessel modes (3.2) have infinite energy and, when propagated in a homogeneous space, retain their intensity (square of the magnitude modulus (3.2)), and therefore are called Bessel diffraction-free beams [1]. The linear combination of solutions (3.2) of the equation (3.1) with arbitrary coefficients is also a solution of equation (1). In [115], an algorithm was proposed for calculating the phase optical element, which forms Bessel diffraction-free beams with a given mode composition:

$$E(r, \phi, z = 0) = \sum_{n=0}^{\infty} C_n \exp(in\phi) J_n(kr\sin\theta). \tag{3.3}$$

In [116] it was proposed to consider the Mathieu beam as an alternative to the Bessel beam:

$$E_n(\xi,\eta,z) = Ce_n(\xi,q)ce_n(\eta,q)\exp(ikz\cos\theta_0), \tag{3.4}$$

where $Ce_n(\xi, q)$, $ce_n(\eta, q)$ are the non-periodic and periodic Mathieu functions. Solution (3.4) is a solution [114] of equation (3.1) in an elliptical coordinate system:

$$
\begin{aligned}
x &= d\operatorname{ch}\xi\cos\eta, \\
y &= d\operatorname{sh}\xi\sin\eta, \\
z &= z, \\
q &= (kd\sin\theta_0)^2 / 4.
\end{aligned}
\tag{3.5}
$$

In [117] it is shown that the linear combination of the even and odd Mathieu beams

$$E_n(\xi,\eta,z) = Ce_n(\xi,q)ce_n(\eta,q) + iSe_n(\xi,q)se_n(\eta,q), \quad n\geq 2 \tag{3.6}$$

has n isolated zeros with a unit topological charge of the same sign, lying on the horizontal x axis inside the first bright elliptical ring. That is, the linear combination (3.6) of the Mathieu beams (3.4), which do not have an orbital angular momentum (OAM), has an OAM. It is interesting [110] that a linear combination, analogous to (3.6), of two Hermite–Gauss modes that do not possess an OAM, possesses an OAM.

The periodic Mathieu functions can be decomposed into a Fourier series [114], for example, the even functions:

$$ce_{2n}(\phi,q) = \sum_{m=0}^{\infty} A_{2m}^{2n(q)}\cos(2m\phi), \tag{3.7}$$

the coefficients of the series (3.7) are calculated recurrently. Using (3.7), the Mathieu diffraction-free beam (3.4) can be represented as a linear combination of Bessel modes:

$$E_{2n}(\xi,\eta,z) = \exp(ikz\cos\theta_0)\sum_{m=0}^{\infty}(-1)^m A_{2m}^{2n}\cos(2m\phi)J_{2m}(kr\sin\theta_0), \tag{3.8}$$

where (ξ, η, z) and (r, φ, z) are the elliptical and cylindrical coordinates. Work [118] considered diffraction-free beams, described as a linear combination of Bessel modes (3.8). In fact, the beams (3. 8) are Mathieu beams (3.4) in the cylindrical coordinates.

It is interesting to find linear combinations of the Bessel modes (3. 3) or (3. 8), which would be described by simple analytic functions, with which one could analytically calculate some properties of such diffraction-free beams. For example, in this section we consider a linear combination of Bessel modes (3.2) with coefficients such that the series (3.3) is calculated and equal to the Bessel function with a complex argument.

It is shown that such a diffraction-free elegant Bessel beam (EB-beam) has a countable number of isolated intensity zeros located on the x axis near the zeros of the Bessel mode of the lowest order in a linear combination. All these zeros (except the axial one) generate optical vortices with a single topological charge and opposite signs from different sides from the origin. The zero intensity on the optical axis generates an optical vortex with a topological charge n. The OAM of such beams was precisely calculated and it turned out to be fractional.

3.1.1. Zeros of asymmetric Bessel beams

Consider the following superposition of the Bessel modes:

$$E_n\left(r,\varphi,z=0\right)=\sum_{p=0}^{\infty}\frac{\exp\left(in\varphi+ip\varphi\right)}{p!}J_{n+p}\left(\alpha r\right). \tag{3.9}$$

The field (3.9) forms the Bessel diffraction-free mode for any integer n. Interestingly, the amplitude (3.9) does not explicitly depend on the wavelength λ, but only through the parameter $\alpha = k\sin\theta_0 = (2\pi/\lambda)\sin\theta_0$. That is, for any wavelength, one can choose the angle θ_0 so that the parameter α does not change and the Bessel beam (3.9) remains the same. In an arbitrary plane, transverse to the optical axis z, such beams will have the following complex amplitude:

$$E_n\left(r,\varphi,z\right)=\exp\left(i\sqrt{k^2-\alpha^2}\,z\right)\sum_{p=0}^{\infty}\frac{\exp\left(in\varphi+ip\varphi\right)}{p!}J_{n+p}\left(\alpha r\right). \tag{3.10}$$

We consider a linear combination of the Bessel modes in the form of (3.10), because this series is equal to the Bessel function with a complex argument. Indeed, in [65] there is a reference series (expression 5.7.6.1):

$$\sum_{k=0}^{\infty} \frac{t^k}{k!} J_{k+v}(x) = x^{v/2} (x-2t)^{-v/2} J_v\left(\sqrt{x^2-2tx}\right). \qquad (3.11)$$

With the help of (3.11) we transform (3.9):

$$E_n(r,\varphi,z=0) = \sum_{p=0}^{\infty} \frac{\exp(in\varphi + ip\varphi)}{p!} J_{n+p}(\alpha r) =$$

$$= \left[\frac{\alpha r}{\alpha r - 2\exp(i\varphi)}\right]^{n/2} J_n\left\{\sqrt{\alpha r\left[\alpha r - 2\exp(i\varphi)\right]}\right\} \exp(in\varphi). \qquad (3.12)$$

The Bessel function on the right side (3.12) will not diverge at infinity, since all the exponents and the Bessel functions on the left side (3.12) do not exceed one in absolute value, which means that the entire series does not exceed the number e. When the denominator in (3.12) is zero, then the argument of the Bessel function is zero. The zero-to-zero uncertainty expands, and the entire expression (3.12) is zero (for $n > 0$). The zeros of the EB-beam (3.12) lie on the x axis (for $\varphi = \pi m$ where $m = 0, 1, 2,...$) in points with the coordinates:

$$x_{+p} = \frac{1+\sqrt{1+\gamma_p^2}}{\alpha}, \quad x > 0;$$

$$x_{-p} = \frac{1-\sqrt{1+\gamma_p^2}}{\alpha}, \quad x < 0, \quad p = 1, 2, 3,... \qquad (3.13)$$

In (3.13) γ_p are the roots of the Bessel function: $J_n(\gamma_p) = 0$. From (3.13) it follows that the intensity in the cross section of the EB-beam is asymmetric with respect to the origin, since $x_{+p} > -x_{-p}$. All zeros of the EB-beam (except axial with $r = 0$) generate optical vortices with a single topological charge and opposite signs from different sides from the origin. At $x < 0$ the topological charge of optical vortices is $+1$, for $x > 0$ the topological charge is -1. Indeed, consider the expression under the root in the argument of the Bessel function in (3.12). If root $x = \gamma_p > 2/\alpha$, then at the beginning of a detour around this zero, the polar angle changes in the range $0 < \varphi < \pi/2$ and under the root in the Bessel function argument in (3.12) we get: $(\alpha r)^2 - 2(\alpha r)\cos \varphi - i2(\alpha r)\sin \varphi = x - iy$ (optical vortex with topological charge -1). If root $x = \gamma_p < 0$, then at the beginning of the traverse around this zero the polar angle can be written as

$\varphi = \pi + \psi$, wherein ψ varies in the range $0 < \psi < \pi/2$, and under the root in the argument of the Bessel function (3.12) we obtain $(ar)^2 + 2(ar)\cos \psi - i2(ar)\sin \psi = x - iy$ (the optical vortex with the topological charge +1).

To change the signs of these optical vortices to opposite ones, it is necessary instead of (3.12) to take a complex conjugate expression. Zero intensity in (3.12) on the optical axis generates an optical vortex with topological charge n.

These conclusions can be checked by examining the phases in Figs. 3.1 and 3.2 which show for different values of the parameters the intensities of the light beams (3.12) in the initial plane, as well as the intensity cross sections at $z = y = 0$ and $z = x = 0$. In the simulation, the following parameter values were used: the wavelength of light $\lambda = 532$ nm, scaling factor $\alpha = 0.2$ (μm)$^{-1}$, the computational domain boundaries $-150\,\lambda \le x, y \le 150\,\lambda$, the number of counts for each axis: $N = 200$. The order of the beam in Figs. 3.1 and 3.2 is equal to $n = 0, 3$.

The zero-order EB-beam (Fig. 3.1) has an interesting property: it has a maximum intensity and a non-zero OAM (it will be calculated

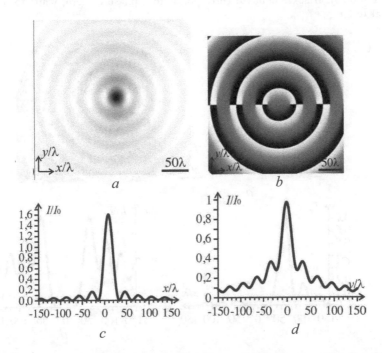

Fig. 3.1. Intensity (a) and phase (b) of a light beam (3.12) of zero order ($n = 0$) in the initial plane, as well as the intensity cross section at $z = y = 0$ (c) and $z = x = 0$ (d). Black colour (b) corresponds to a phase of zero, white to the phase equal to 2π.

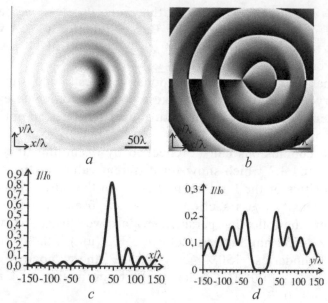

Fig. 3.2. Intensity (*a*) and phase (*b*) of the light beam (3.12) of the third order ($n = 3$) in the initial plane, as well as the intensity cross section at $z = y = 0$ (*c*) and $z = x = 0$ (*d*). Black colour (*b*) corresponds to a phase of zero, white to the phase equal to 2π.

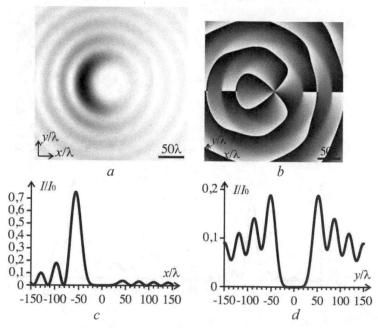

Fig. 3.3. Intensity (*a*) and phase (*b*) of a fourth-order light beam (3.16) ($n = 4$) in the initial plane, as well as the intensity cross section at $z = y = 0$ (*c*) and $z = x = 0$ (*d*). Black colour (*b*) corresponds to a phase of zero, white to the phase equal to 2π.

below). This property can be used when manipulating dielectric microparticles. A particle that is several times larger than the main maximum of the beam intensity in Fig. 3.1, can be kept by this intensity maximum and, simultaneously, rotate around its axis.

Figure 3.1 shows that the maximum value of a zero order EB-beam

$$E_0(r,\phi) = J_0\left\{\sqrt{\alpha r[\alpha r - 2\exp(i\phi)]}\right\} \tag{3.14}$$

is more than one: I_{max} = 1.6. There is an explanation. Since when $\alpha r < 2$ and $\varphi = 0$, the argument (3.14) is purely imaginary, then on this section of the x - axis, a zero-order EB-beam can be represented as

$$E_0(r,\phi) = I_0\left\{\sqrt{\alpha x(2-\alpha x)}\right\}, \tag{3.15}$$

where $I_0(x)$ is the modified zero-order Bessel function. At the ends of the segment $[0, 2/\alpha]$, the argument of the function (3.15) takes the value zero, and the Bessel function itself is $J_0(0) = I_0(0) = 1$. On this segment $[0, 2/\alpha]$, function (3.15) reaches a maximum at the point at which the argument reaches a maximum. And the argument reaches a maximum in the middle of the segment, that is, at $x = 1/\alpha$. At this point, function (3.15) is equal to $E_0(r = 1/\alpha, \varphi = 0) = I_0(1) \cong 1.266$. Therefore, the maximum intensity of the EB-beam is $I_{max} = |E_0(r = 1/\alpha, \varphi = 0)|^2 = I_0^2(1) \cong 1.60$. From this, in particular, it follows that the maximum intensity of the side lobes of a zero-order EB-beam (3.14) (Fig. 3.1), which coincides with the maximum intensity of the side lobes for a typical Bessel beam $J_0(x)$, is a lower part of the maximum intensity in the main lobe for the EB-beam (10%) than for the ordinary Bessel beam (16%).

Figure 3.2 shows that the main lobe of a high-order EB-beam has the form of a 'crescent', convex to the right. An expression can be written for an EB-beam having the form of a 'crescent', convex to the left. The EB-beam having an intensity distribution, mirror-symmetrical about the axis $x = 0$ compared with the beam (3.12), has an amplitude of the form:

$$E_n(r,\phi) = \sum_{p=0}^{\infty} \frac{(-1)^p \exp[i(n+p)\phi]}{p!} J_{n+p}(\alpha r) =$$

$$= \left[\frac{\alpha r}{\alpha r + 2\exp(i\phi)}\right]^{n/2} J_n\left\{\sqrt{\alpha r[\alpha r + 2\exp(i\phi)]}\right\} \exp(in\phi). \tag{3.16}$$

Figure 3.3 shows, for the same parameter values as in Figs. 3.1 and 3.2, the intensity and phase of the light beam (3.16) in the initial plane, as well as the intensity cross section at $z = y = 0$ and $z = x = 0$. The beam order in Fig. 3.3 is equal to $n = 4$.

3.1.2. Angular spectrum of plane waves of an EB-beam

Substitute the angular spectrum of the Bessel beam $F_n(\theta,\varphi) = (-i)^n \exp(in\ \varphi)\delta(\theta-\theta_0)$ in each term of the series on the left side (3.12), we get:

$$A(\theta,\phi) = \sum_{p=0}^{\infty} \frac{(-i)^p \exp[i(n+p)\phi]}{p!} \delta(\theta-\theta_0) =$$

$$= \frac{(-i)^n}{2\pi \sin\theta_0} \exp[in\phi - i\exp(i\phi)]\delta(\theta-\theta_0). \tag{3.17}$$

Indeed, with the substitution of (3.17) into the expansion in plane waves in cylindrical coordinates (r, ξ, z)

$$E_n(r,\xi,z) = \int_0^{2\pi} \int_{-\pi}^{\pi} A(\theta,\phi)\exp \times$$

$$\times \left[ikr\cos(\phi-\xi)\sin\theta + ikz\cos\theta\right]\sin\theta d\theta d\phi, \tag{3.18}$$

is obtained (3.12).

3.1.3. The orbital angular momentum of the EB-beam

The orbital angular momentum J_z (the projection of the OAM on the optical axis) and the total intensity I of the light beam in the plane transverse to the optical axis are determined by formulas [110]:

$$J_z = \text{Im}\left\{ \iint_{\mathbb{R}^2} E^* \frac{\partial E}{\partial \varphi} r dr d\varphi \right\} = \text{Im}\left\{ \lim_{R \to \infty} \int_0^R \int_0^{2\pi} E^* \frac{\partial E}{\partial \varphi} r dr d\varphi \right\}, \qquad (3.19)$$

$$I = \iint_{\mathbb{R}^2} E^* E r dr d\varphi = \lim_{R \to \infty} \int_0^R \int_0^{2\pi} E^* E r dr d\varphi. \qquad (3.20)$$

Substituting in (3.19), (3.20) the complex amplitude from the left side of (3.12), we obtain the orbital angular momentum and the total intensity of the EB-beam:

$$J_z = 2\pi \lim_{R \to \infty} \sum_{p=0}^{\infty} \frac{n+p}{(p!)^2} \int_0^R J_{n+p}^2 (\alpha r) r dr, \qquad (3.21)$$

$$I = 2\pi \lim_{x \to \infty} \sum_{p=0}^{\infty} \frac{1}{(p!)^2} \int_0^R J_{n+p}^2 (\alpha r) r dr. \qquad (3.22)$$

Integrals in these expressions are described in [112] (expression 5.54.2):

$$\int J_p^2 (\alpha r) r dr = \frac{r^2}{2} \left[J_p^2 (\alpha r) - J_{p-1} (\alpha r) J_{p+1} (\alpha r) \right]. \qquad (3.23)$$

Using the asymptotic Bessel function for large values of the argument (9.2.1 expression in [53]), we find that all the integrals in (3.21) and (3.22) are independent of the order of the Bessel functions and are $R/(\pi\alpha)$. Then, using the number of the left rows of 0.246.1 and 0.246.2 from [112] and dividing (3.21) by (3.22), we obtain the expression for the orbital angular momentum, normalized by the intensity:

$$\frac{J_z}{I} = n + \frac{I_1(2)}{I_0(2)} \approx n + 0.69777. \qquad (3.24)$$

where $I_n(z)$ is the modified Bessel function.

From (3.24) it follows that the OAM of the EB-beams is fractional and linearly increases with order n. It also follows from (3.24) that a zero-order EB beam also has an OAM equal to $J_z/I = I_1(2)/I_0(2) = 0.69777$. In the numerical calculation (by the formulas (3.19) and (3.20)) it turned out for the parameters $n = 8$, $\lambda = 532$ nm, $\alpha = 0.2$

$(\mu m)^{-1}$, $-150\lambda \leq x$, $y \leq 150\ \lambda$, $N = 400$ that OAM is equal to $J_z/I = 8.667$, and differs from the OAM calculated by the formula (3.24) $J_z/I = 8.6977$, a total of 0.3%. Note that the approach for calculating the OAM used in this section is applicable for calculating the OAM of any light field represented by the superposition of the Bessel modes.

3.1.4. The orthogonality of EB-beams

Just as in the previous section we calculated the orbital angular momentum, we can calculate the scalar product of two EB beams – an nth order beam with the α parameter and an mth order beam with the β parameter:

$$\left(E_{n\alpha}, E_{m\beta}\right) = \int\limits_0^\infty \int\limits_0^{2\pi} E_{n\alpha} E_{m\beta}^* r\, dr\, d\varphi = 2\pi \frac{\delta(\alpha - \beta)}{\alpha} I_{n-m}(2). \qquad (3.25)$$

where $I_{n-m}(x)$ is the modified Bessel function.

From the expression (3.25) it can be seen that, in contrast to the Bessel modes, the EB-beams are orthogonal only by a scaling factor, and the Bessel functions are not orthogonal in order.

The following results were obtained in this section [88]. The two-parameter family of nonparaxial diffraction-free asymmetric elegant Bessel beams described by the Bessel functions of the first kind of integer order with complex arguments (equations (3.12) and (3.16)) is proposed. The asymmetric Bessel beams have a countable number of isolated intensity zeros lying on the horizontal x axis and having a single topological charge (except for the axial intensity zero) and different signs from different sides of the optical axis (equation (3.13)); The axial intensity zero has a topological charge equal to the order of the Bessel function. The asymmetric Bessel beams have a ring spectrum with respect to the angle defining the direction of the wave vector, and only the phase distribution over the polar angle (equation (3.17)). The asymmetric Bessel beams have a fractional OAM, which grows linearly with an increase in the mode number (equation (3.24)); a zero-order beam has an intensity maximum on the optical axis and isolated intensity zeros near the roots of the Bessel function and has an OAM of $0.69777\hbar$ per photon. These beams are orthogonal in the continuous parameter and not orthogonal on an integer parameter (equation (3.25)).

3.2. Asymmetric Bessel–Gauss beams

In 1987, Gori [119] considered Bessel–Gauss beams (BG-beams). The complex amplitude of such beams is described by the product of the Gaussian function by the nth order Bessel function of the first kind and the phase function describing the angular harmonic. The complex amplitude of the BG-beam satisfies the paraxial propagation equation. These beams have a radially symmetric intensity distribution and have an orbital angular momentum (OAM). Radial symmetry is preserved when the beam propagates. But BG-beams are not free-space modes, since with propagation not only scale changes but also the light energy is redistributed between different rings in the intensity distribution in the beam cross section. The BG-beams have been summarized in several papers [120–122]. For example, in [122] the Helmholtz–Gauss beams a special case of which are the Bessel–Gauss beams are considered [119]. The BG beams have finite energy, but Bessel modes that do not possess finite energy are also known [1]. The Bessel modes are solutions to the Helmholtz equation [114] and, when propagated in a homogeneous space, retain their intensity, and therefore are also called diffracion-free Bessel beams [1]. The linear combination of the Bessel modes with arbitrary coefficients is also a solution of the Helmholtz equation. In [115], an algorithm was proposed for calculating the phase optical element which forms the Bessel diffraction-free beams with a given mode composition. In [116] it was proposed to consider the Mathieu beam as an alternative to the Bessel beam. In [117] it was shown that a linear combination of the even and odd Mathieu beams with complex coefficients has an OAM. But the Mathieu beams themselves do not possess OAM. It is interesting [110] that a linear combination of two Hermite–Gauss modes with complex coefficients that do not possess an OAM, possesses an OAM. The periodic Mathieu function can be decomposed into a Fourier series [114]. For example, the Mathieu even functions are expanded in cosines from the polar angle in a cylindrical coordinate system, and the odd functions in sine terms. Therefore, the Mathieu diffraction-free beam can be represented as a linear combination of the Bessel modes. Such beams are considered in [118].

It is interesting to find linear combinations of the BG-beams, which would be described by simple analytic functions, with which one could analytically calculate some properties of such beams. In this section, we consider a linear combination of the BG-beams, which is described by the Bessel function with a complex argument. It is shown that such an asymmetric Bessel–Gauss beam (aBG beam) has in the initial plane a countable number of isolated intensity zeros located on the horizontal axis. All these zeros (except for the axial one) generate optical vortices with a single topological charge and opposite signs from different sides from the origin. The zero intensity on the optical axis generates an optical vortex with topological charge n. When propagating in free space, the aBG beams rotate around the optical axis. It is shown that such beams have an OAM, which increases with the number n, and also increases with an increase in the asymmetry parameter c of the beam. Moreover, the OAM of aBG beams can be either integer or fractional.

3.2.1. Linear combination of BG-beams

We write the complex amplitude of the BG-beam [119] in the initial plane $z = 0$:

$$E_n\left(r,\varphi,z=0\right)=\exp\left(-\frac{r^2}{\omega_0^2}+in\varphi\right)J_n\left(\alpha r\right), \qquad (3.26)$$

where $\alpha = k \sin \theta_0 = (2\pi/\lambda) \sin\theta_0$ is the scaling factor, $k = 2\pi/\lambda$ is the wavenumber of light with the wavelength λ, θ_0 is the angle of the conical wave forming the Bessel beam. In any other z plane, the complex amplitude (3.26) will be:

$$E_n(r,\varphi,z)=q^{-1}\left(z\right)\exp\left(ikz-\frac{i\alpha^2 z}{2kq(z)}\right)=$$
$$=\exp\left(-\frac{r^2}{\omega_0^2 q(z)}+in\varphi\right)J_n\left[\frac{\alpha r}{q(z)}\right], \qquad (3.27)$$

where $q(z) = 1+iz/z_0$, $z_0 = k\omega_0^2/2$ is the Rayleigh length, ω_0 is the waist radius of the Gaussian beam, $J_n(x)$ is the Bessel function of the first kind of nth order. The beams (3.27) are not paraxial modes of free space, since the argument of the Bessel function is complex. Consider the following superposition of the BG-beams $q = q(z)$:

$$E_n(r,\varphi,z;c) = q^{-1} \exp\left(ikz - \frac{i\alpha^2 z}{2kq} - \frac{r^2}{q\omega_0^2} \right) \times$$

$$\times \sum_{p=0}^{\infty} \frac{c^p \exp(in\varphi + ip\varphi)}{p!} J_{n+p}\left(\frac{\alpha r}{q} \right).$$
(3.28)

The field (3.28) forms a paraxial asymmetric Bessel–Gauss beam (aBG beam) for any integer n and any complex constant c. But for simplicity, in the following, we will consider a constant with a real positive value: $c \geq 0$. Although, if we consider c as a complex value $c = |c| \arg c$ or consider that it may be negative $c < 0$, then the distribution of the field intensity (3.28) will turn through the angle $\arg c$ around the optical axis. With $c = 0$ in (3.28) only one item remains non-zero with $p = 0$, and the aBG beam becomes the usual BG beam (3.26), (3.27).

Here we consider the linear combination of the BG-beams in the form (3.28) because this series is equal to the Bessel function with a complex argument. Indeed, in [65] there is a reference series (expression 5.7.6.1):

$$\sum_{k=0}^{\infty} \frac{t^k}{k!} J_{k+v}(x) = x^{v/2} (x - 2t)^{-v/2} J_v\left(\sqrt{x^2 - 2tx} \right).$$
(3.29)

With the help of (3.29) we transform (3.28) and get:

$$E_n(r,\varphi,z;c) = \frac{1}{q} \exp\left(ikz - \frac{i\alpha^2 z}{2kq} - \frac{r^2}{q\omega_0^2} + in\varphi \right) \times$$

$$\times \left[\frac{\alpha r}{\alpha r - 2cq \exp(i\varphi)} \right]^{n/2} \times J_n\left\{ q^{-1} \sqrt{\alpha r [\alpha r - 2cq \exp(i\varphi)]} \right\}.$$
(3.30)

The expression (3.30) is a closed form for the complex amplitude of the three-parameter family of the paraxial scalar aBG beams. Two continuous real parameters of the aBG beams control the scale (α) and the degree of asymmetry (c). The radius of the Gaussian beam ω_0 is assumed to be the same for the whole family of the aBG beams. The integer parameter n together with continuous c determine the magnitude of the OAM. When the denominator in (3.30) is zero, then the argument of the Bessel function is zero. Zero to zero uncertainty is revealed.

A beam (3.30) has a countable number of isolated intensity zeros, which generate optical vortices with a single topological charge, except for the axial zero, which generates an optical vortex with the nth topological charge. To obtain the polar coordinates of the aBG beam, we equate the argument of the Bessel function in (3.30) to the root of the Bessel function $\gamma_{np}(J_n(\gamma_{np}) = 0)$:

$$\alpha^2 r^2 - 2\alpha c q r \exp(i\varphi) = \gamma_{np}^2 q^2. \tag{3.31}$$

Select the real and imaginary parts of this equation:

$$\begin{cases} \alpha^2 r^2 - 2\alpha c |q| r \cos(\varphi + \Psi) = \gamma_{np}^2 |q|^2 \cos(2\Psi), \\ -2\alpha c |q| r \sin(\varphi + \Psi) = \gamma_{np}^2 |q|^2 \sin(2\Psi), \end{cases} \tag{3.32}$$

wherein $\Psi = \arctan(z/z_0)$ is the Gouy phase. From (3.32) we find the coordinates of points with zero intensity:

$$\begin{cases} \varphi_{np} = \dfrac{1}{2}\arccos\left[\cos(2\Psi) - \dfrac{\gamma_{np}^2}{2c^2}\sin^2(2\Psi)\right], \\ r_{np} = \dfrac{|q|}{\alpha}\sqrt{\gamma_{np}^2 \cos(2\Psi) + 2c^2 \pm 2\sqrt{D}}, \end{cases} \tag{3.33}$$

where $D = \left(c^2 - \gamma_{np}^2 \sin^2 \Psi\right)\left(c^2 + \gamma_{np}^2 \cos^2 \Psi\right)$. From both equations (3.33) it follows that in order for the coordinates of the zeros to be real, the following condition must be met:

$$\gamma_{np} \sin \Psi \le c. \tag{3.34}$$

From (3.34) it follows that if $c > \gamma_{np}$, then all intensity zeros with numbers from $q = 0$ to $q = p$ do not disappear during propagation (since for these zeros $\gamma_{nq} < c$ and inequality (3.34) holds for any distances z), but the rest zeros with numbers $q = p + 1, p + 2,\ldots$ (for which $c \le \gamma_{nq}$) disappear at some distance $z = cz_0/(\gamma_{nq}^2 - c^2)^{1/2}$.

When $z = 0$ instead of (3.33) we write:

$$\begin{cases} r_{p+} = \alpha^{-1}\sqrt{c + \sqrt{c^2 + \gamma_{np}^2}}, & \varphi_{np} = 2p\pi, \\ r_{p-} = \alpha^{-1}\sqrt{\sqrt{c^2 + \gamma_{np}^2} - c}, & \varphi_{np} = (2p+1)\pi. \end{cases} \tag{3.35}$$

From (3.35) it follows that the intensity in the cross section of the aBG beam is asymmetric with respect to the origin, since $r_{p+} > r_{p-}$. And when $c > 0$ increases, the asymmetry of the aBG beam increases, and at $c = 0$ the isolated zeros of the aBG beam disappear, and rings of zero intensity appear in the radially symmetric Bessel– Gauss beam. All zeros of the aBG beam (except for the axial beam at $r = 0$) generate optical vortices with a single topological charge and opposite signs from different sides from the origin. For intensity zeros with radial coordinates r_{p+}, the topological charge of optical vortices is +1, for r_{p-} the topological charge is −1. In order to change the signs of these optical vortices to the opposite ones, it is necessary instead of (3.30) to take a complex conjugate expression. From (3.33) it can be seen that the intensity zeros, and hence the intensity distribution itself in the transverse plane of the aBG beam, rotate during propagation. But the rotation is quite complicated. Only when $c \gg 1$, in the first equation in (3.33) the second term in square brackets can be neglected, and then the intensity pattern in the beam cross section rotates as a whole. In this case, the polar angle changes with distance z:

$$\varphi = \arctan\left(\frac{z}{z_0}\right). \tag{3.36}$$

From (3.36) it follows that the aBG beam from the initial plane $z = 0$ to the Rayleigh distance $z = z_0$ will rotate counterclockwise by $\pi/4$, and for the rest of the path from $z = z_0$ to $z = \infty$ by another $\pi/4$, and thus the beam will rotate $\pi/2$ all the way. The rotation of the aBG beam does not depend on the number n, that is, the zero beam will also rotate at $n = 0$.

Figure 3.4 shows the distribution of both the intensities and phases of an aBG beam (3.30) for different values of the asymmetry parameter c. In the simulation the following parameters were used: wavelength $\lambda = 532$ nm, a Gaussian beam waist radius $\omega_0 = 10\lambda$, the scaling factor $\alpha = 1/(10\lambda)$, the boundary of the computational domain $-40\lambda \leq x, y \leq 40 \lambda$. Figure 3.4 shows that with increasing $c > 0$ intensity from the ring form becomes like a crescent, bulging to the right. One can write an expression for an aBG beam, having the form of a crescent, convex to the left. The aBG beam, which has an intensity distribution that is mirror-symmetric about the $x = 0$ axis compared to the beam (3.30), has an amplitude of the form:

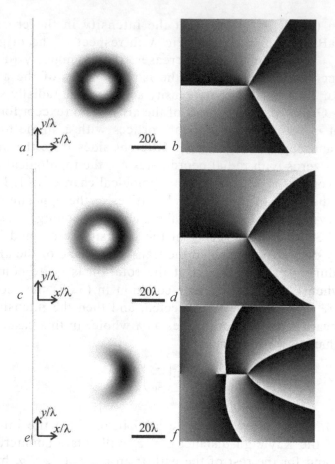

Fig. 3.4. The intensity distribution (negative) (*a, c, e*) and phase (*b, d, f*) of the third-order light beam (3.30) (*n* = 3) in the initial plane at z = 0 for different values of the asymmetry parameter *c*: 0.1 (*a, b*); 1 (*c, d*); 10 (*e, f*).

$$E_n(r,\varphi,z=0;c) = \exp\left(-\frac{r^2}{\omega_0^2}\right) \times$$

$$\times \sum_{p=0}^{\infty} \frac{(-c)^p \exp[i(n+p)\varphi]}{p!} J_{n+p}(\alpha r) = \exp\left(-\frac{r^2}{\omega_0^2}\right) \times \tag{3.37}$$

$$\times \left[\frac{\alpha r}{\alpha r + 2c\exp(i\varphi)}\right]^{n/2} J_n\left\{\sqrt{\alpha r[\alpha r + 2c\exp(i\varphi)]}\right\} \exp(in\varphi).$$

Also in Fig. 3.4 it is seen that the intensity zeros lying on the x < 0 axis with increasing *c* approach the origin of coordinates: in Fig. 3.4 *b, d* they are not yet visible, and in Fig. 3.4 *f* one zero

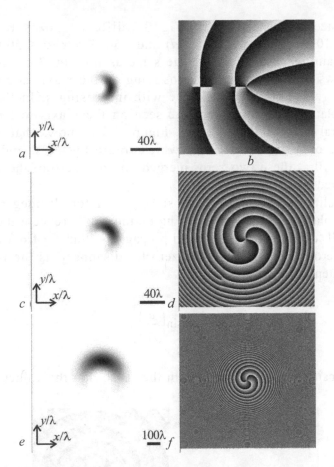

Fig. 3.5. The intensity distribution (negative) (a, c, e) and phase (b, d, f) of the third-order beam (3.30) of the third order ($n = 3$) at different distances: $z = 0$ (a, b), $z = z_0$ (c, d), $z = 10z_0$ (e, f).

appeared. This behaviour of zeros follows from the second equation (3.35) for large $c \ll 1$. In the centre of Fig. 3.4 b, d, f is located the zero intensity of the third order ($n = 3$).

Figure 3.5 shows the intensity and phase distributions of a third-order aBG beam ($n = 3$) with a relatively large value of the parameter $c = 10$ at different distances along the optical axis in the $z = 0$ (a, b), $z = z_0$ (c, d) planes, $z = 10z_0$ (e, f). Figure 3.5 shows that the beam rotates around the optical axis. At the Rayleigh distance z_0 the intensity picture turned counterclockwise by 45 degrees (Fig. 3.5 c), and at 10 z_0 the intensity picture turned almost 90 degrees (Fig. 3.5 e).

The size of the pictures in Fig. 3.5 differs: $-80\lambda \leq x, y \leq 80\lambda$ (a, b); $-100\lambda \leq x, y \leq 100\lambda$ (c, d) and $-500\lambda \leq x, y \leq 500\lambda$ (e, f). The remaining parameters are the same as for Fig. 3.4. Figure 3.5 also shows that the intensity zeros lying on the x axis at $z = 0$ also begin to rotate counterclockwise with increasing z. In Fig. 3.5 b three isolated zero intensities are seen on the x axis to the left of the central zero of the 3rd order. Figure 3.5 d shows that at $z = z_0$ there are only two zeros left, which rotate by $45°$s, and in Fig. 3.5 f $z = 10$ z_0 these two zeros merged into one zero of the 2nd order and rotated by almost $90°$.

Inequality (3.34) implies that, since $\gamma_{n1} > 1$ for all integer n, for c ≤ 1 all isolated intensity zeros lying on the x axis (except the central zero) will 'disappear' as the beam propagates, starting from the most distant zeros ($\gamma_p \gg 1$). The last zero to 'disappear' is the first zero of the intensity γ_1 at z equal to:

$$z = z_0 \tan\left[\arcsin\left(\frac{c}{\gamma_1}\right)\right]. \tag{3.38}$$

The intensity zeros «disappear» in the sense that the values of their

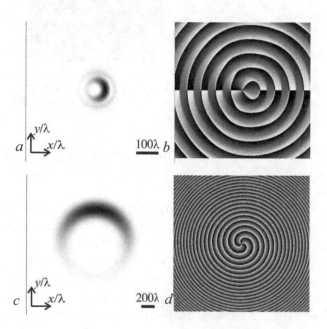

Fig. 3.6. The intensity distribution (negative) (a, c) and phase (b, d) of a third-order aBG beam $(n = 3)$ with $c = 1$ at different distances z: 0 (a, b) and z_0 (c, d).

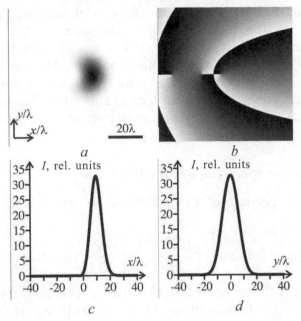

Fig. 3.7. Intensity (negative) (*a*) and phases (*b*) of a light beam (3.30) of zero order ($n = 0$) in the initial plane, as well as intensity cross sections at $z = y = 0$ (*c*) and $z = x = 0$ (*d*).

coordinates, according to (3.33) and under condition (3.34), become complex, and not real.

Figure 3.6 shows the intensity distribution and phase of the aBG beam with a small asymmetry parameter $c = 1$. The other calculation parameters: wavelength $\lambda = 532$ nm, Gaussian beam waist radius $\omega_0 = 100\lambda$, the scaling factor $\alpha = 1/(10\lambda)$, the boundary of the computational domain $-300\lambda \leq x, y \leq 300\lambda$ (Fig. 3.6 *a*, *b*) and $-1000\lambda \leq x, y \leq 1\,000\lambda$ (Fig. 3.6 *c*, *d*). With these parameters, the rotation of the aBG beam is no longer described by the simple formula (3.36), since the asymmetry parameter c is small. Figure 3.6 shows that the intensity pattern rotates at a distance $z = z_0$ by almost 90° counterclockwise. And the isolated intensity zeros that lie on the x axis and are visible in Fig. 3.6 *b*, 'disappear' during the propagation of the beam: in Fig. 3.6 *d* at $z = z_0$ there are no the intensity zeros, except the central one.

Figure 3.7 shows the intensity and phase of the aBG beam of the zeroth order with the following parameters: wavelength $\lambda = 532$ nm, the waist radius of the Gaussian beam $\omega_0 = 10\lambda$, the scaling factor

$\alpha = 1/(10\lambda)$, the asymmetry parameter $c = 10$, the calculated domain $-40\lambda \leq x, y \leq 40\lambda$. In the case of a zero order the aBG beam (Fig. 3.7) has an interesting property: it has a maximum intensity near the optical axis and the OAM other than zero. This property can be used when manipulating dielectric microparticles. A particle that is several times larger than the main maximum of the beam intensity in Fig. 3.7, can be kept by this maximum intensity and, simultaneously, rotate around its axis.

3.2.2. Fourier spectrum of an aBG beam

The angular spectrum of a conventional BG beam is known [122]:

$$A_n(\rho,\phi) = (-i)^n \exp\left[-\left(\frac{k\rho\omega_0}{2f}\right)^2\right] I_n\left(\frac{k\alpha\rho\omega_0^2}{2f}\right) \exp(in\phi), \qquad (3.39)$$

where $I_n(x)$ is the nth order modified Bessel function, ρ is the radial coordinate in the Fourier plane, f is the focal length of the spherical lens forming the spatial spectrum of the BG beam. Consider a linear combination similar to (3.28), but consisting of the functions (3.39), we obtain the angular spectrum of an aBG beam:

$$A_n(\rho,\phi) = \exp\left[-\left(\frac{k\rho\omega_0}{2f}\right)^2\right] \sum_{p=0}^{\infty} \frac{(-ic)^p \exp\left[i(n+p)\phi\right]}{p!} I_p\left(\frac{k\alpha\rho\omega_0^2}{2f}\right). \qquad (3.40)$$

Using the reference ratio [65]:

$$\sum_{k=0}^{\infty} \frac{t^k}{k!} I_{k+v}(x) = x^{v/2}(x+2t)^{-v/2} I_v\left(\sqrt{x^2 + 2tx}\right), \qquad (3.41)$$

we obtain the final expression for the Fourier spectrum of the aBG beam:

$$A_n(\rho,\phi) = \exp\left[-\left(\frac{k\rho\omega_0}{2f}\right)^2 + in\phi\right] \times$$

$$\times \left(\frac{\xi}{\xi + 2ce^{i(\phi-\pi/2)}}\right)^{n/2} I_p\left\{\sqrt{\xi\left[\xi + 2ce^{i(\phi-\pi/2)}\right]}\right\}, \qquad (3.42)$$

where $\xi = \alpha k \rho \omega_0^2 / (2f)$. The angular spectrum (3.42) is asymmetric: at $\varphi = \pi/2$ and on the ring of a fixed radius $\rho = \rho_0$ the amplitude modulus (3.42) has a maximum, and at $\varphi = -\pi/2$ – a minimum. The type of the angular spectrum (3.42) is similar in appearance to the amplitude of the aBG beam (3.30) (up to the replacement of the Bessel function by the modified Bessel function), but rotated by 90 degrees. Therefore, the asymmetry of the spectrum (3.42) will be similar to the asymmetry of the aBG beam (3.30).

3.2.3. The orbital angular momentum of an aBG beam

The orbital angular momentum J_z (the projection of the OAM on the optical axis) and the total intensity I of the light beam in the plane transverse to the optical axis are determined by formula [114]:

$$J_z = \mathrm{Im}\left\{ \iint_{\mathbb{R}^2} E^* \frac{\partial E}{\partial \varphi} r dr d\varphi \right\}, \tag{3.43}$$

$$I = \iint_{\mathbb{R}^2} E^* E r dr d\varphi \tag{3.44}$$

Substituting in (3.43), (3.44) the complex amplitude (3.28) at $z = 0$, we obtain the orbital angular momentum and the total intensity of the aBG beam:

$$J_z = 2\pi \sum_{p=0}^{\infty} \frac{c^{2p}(n+p)}{(p!)^2} \int_0^{\infty} \exp\left(-\frac{2r^2}{\omega_0^2}\right) J_{n+p}^2(\alpha r) r dr, \tag{3.45}$$

$$I = 2\pi \sum_{p=0}^{\infty} \frac{c^{2p}}{(p!)^2} \int_0^{\infty} \exp\left(-\frac{2r^2}{\omega_0^2}\right) J_{n+p}^2(\alpha r) r dr. \tag{3.46}$$

The integrals in these expressions are described in [65]:

$$\int_0^{\infty} x \exp(-px^2) J_v(bx) J_v(cx) dx = (2p)^{-1} \exp\left(-\frac{b^2+c^2}{4p}\right) I_v\left(\frac{bc}{2p}\right). \tag{3.47}$$

Using the integral (3.47) and dividing (3.45) by (3.46), we obtain the expression for the orbital angular momentum, normalized by the intensity:

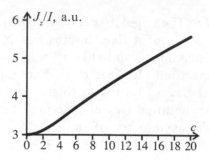

Fig. 3.8. The dependence of OAM asymmetry parameter c at $n = 3$, $\omega_0 = 10\lambda$, $\alpha = 1/(10\lambda)$.

$$\frac{J_z}{I} = n + \sum_{p=0}^{\infty} \frac{c^{2p} p I_{n+p}(y)}{(p!)^2} \left[\sum_{p=0}^{\infty} \frac{c^{2p} I_{n+p}(y)}{(p!)^2} \right]^{-1}, \qquad (3.48)$$

where $y = \alpha^2 \omega_0^2 / 4$. Expression (3.48) cannot be further simplified. From (3.48) it follows that the OAM of the aBG beams is greater than n, as all the terms of the series in (3.48) are positive. That is, as the parameter c grows, the asymmetry of the aBG beam grows and its OAM increases almost linearly, as can be seen from Fig. 3.8. Since the parameters α and c are real positive numbers, the second term in (3.48) can be either an integer or a fractional positive number. Therefore, it follows from (3.48) that a zero-order aBG beam ($n = 0$) can have any OAM. We also note that in (3.48) the scale parameters of the Bessel beam α and the Gaussian beam ω_0 enter as a product, therefore for different aBG beams for which $\alpha\omega_0 = $ const, the OAM will be the same (for equal n and c).

3.2.4. Orthogonality of the family of aBG beams

Just as we calculated the orbital angular momentum, we can calculate the scalar product of complex amplitudes of the two EB beams: the beam of the nth order with parameters α and c and the beam of the mth order with parameters β and d:

$$\left(E_{n\alpha c}, E_{m\beta d} \right) = \frac{\pi \omega_0^2}{2} \left(\frac{d^*}{c} \right)^{\frac{n-m}{2}} \exp\left[-\frac{\omega_0^2}{8} (\alpha^2 + \beta^2) \right] \times$$

$$\times \sum_{p=0}^{\infty} \frac{(cd^*)^{p + \frac{|n-m|}{2}}}{p! (p + |n - m|)!} I_{p + \max(m,n)} \left(\frac{\omega_0^2 \alpha \beta}{4} \right), \qquad (3.49)$$

where $I_v(x)$ is the modified Bessel function. It is clear from expression (3.49) that, unlike the Bessel modes [1], the aBG beams are not orthogonal either by the scaling factor or by the order of the Bessel function, or by the asymmetry parameter. In (3.49), the asymmetry parameters c and d were considered complex, and a^* is the complex conjugation.

We note that aBG beams as the radius of the Gaussian beam tends to infinity $\omega_0 \to \infty$ become Bessel diffraction-free asymmetric elegant modes [123].

3.2.5. Experiment

The experimental setup is shown in Fig. 3.9. The experiment used the PLUTO VIS phase spatial modulator. The light of a solid-state laser with a wavelength of 532 nm is expanded with a collimator, limited to an aperture of 8 mm in diameter. The result is a uniform intensity distribution, which can be considered a plane wave. Then the light passes through the beam-splitting cube and hits the spatial light modulator, is reflected from it and deflected by the beam-splitting cube to the CCD camera. The phase was transmitted to the halftone modulator from a computer to form an aBG beam in combination with a phase of a parabolic lens with a focal length of 960 mm.

The CCD camera moved in a small segment in close proximity to the focus. Figure 3.10 shows the phase distribution (*a*) formed on the modulator (without an added lens) and the intensity distributions recorded by the CCD matrix at a distance of 850 mm (*b*), 900 mm (*c*) and 950 mm (*d*) from the modulator. The dimension of the modulator is 1920×1080 pixels, the size of one sensitive element

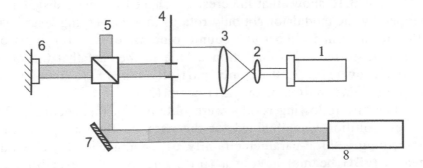

Fig. 3.9. The scheme of the experimental setup. 1 – a solid-state laser with a wavelength of 532 nm, 2, 3 – a collimator, 4 – a diaphragm, 5 – a beam-splitting cube, 6 – PLUTO VIS modulator, 7 – a mirror, 8 – a CCD camera.

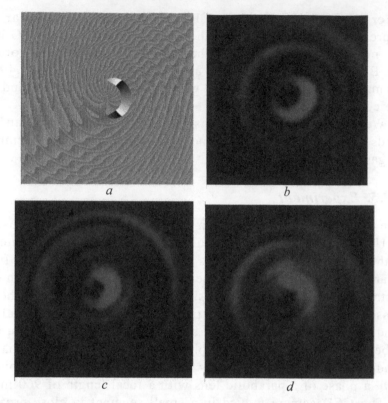

Fig. 3.10. The distribution of the phase formed on the PLUTO VIS modulator (*a*), and the intensity distributions recorded at distances from the modulator: 850 mm (*b*), 900 mm (*c*) and 950 mm (*d*).

is 8 microns. The phase shown in Fig. 3.10, had the dimension of 1024×1024 pixels and was formed in the centre of the modulator. Thus, the exact size of the formed phase distribution was 8.2 mm.

Figurw 3.10 shows that the crescent-shaped intensity distribution formed by the modulator not only rotates in a converging laser beam (the beam turned almost 90° counterclockwise over 100 mm) but also is distorted. This distortion is due to the fact that the phase distribution (Fig. 3.10 *a*) only partially takes into account the amplitude distribution of the aBG beam (3.30).

Thus, the following results were obtained in this section [124]. A new solution was obtained for the paraxial Helmholtz equation describing a three-parameter family of asymmetric Bessel–Gauss beams (aBG beams) having a finite energy and describing the functions of the first kind Bessel integer order with complex argument; when $c = 0$, the aBG beams coincide with the usual Bessel–Gauss beams [119]. The intensity distribution of the aBG beams has a countable number of isolated zeros lying on a horizontal straight line in the initial plane ($z = y = 0$). In the places of these zeros there are optical vortices having single topological charges and different signs from different sides from the optical axis (from the centre of coordinates); The axial intensity zero has a topological charge equal to the order of the Bessel function. The asymmetric Bessel–Gauss beams with large asymmetry parameters $c \gg 1$ rotate as a single unit when distributed in free space: at a distance of Rayleigh's length, they rotate 45°, and another 45° for the rest of the way; with small c the beams rotate in a more complicated way: at a certain distance from the initial plane, optical vortices (isolated intensity zeros) that were present in the beam section disappear (except for the central intensity zero) and at a distance of the Rayleigh length the beam can turn almost 90°. The aBG beams have an asymmetric angular spectrum, expressed in terms of the product of the Gaussian function and the modified Bessel function with a complex argument; The form of this spectrum has an asymmetry similar to the asymmetry of the aBG beam, but rotated 90°. The aBG beams can have an integer and fractional OAM, which grows with increasing number n and almost linearly with increasing parameter c; a zero-order beam has an intensity maximum shifted from the optical axis by c/α and isolated intensity zeros, and can have any OAM, depending on the choice of the asymmetry parameter c. The aBG beams are not orthogonal by any of their parameters n, α, c.

Experimentally using the liquid-crystal light modulator a converging a laser beam was formed, similar to the aBG beam with an intensity distribution in the form of a crescent, which is turned by about 90° counterclockwise.

3.3. Superposition of Bessel modes with complex shift

The Bessel beams, discovered in 1987 [1,125], have many remarkable properties. They propagate without diffraction at a finite distance in free space [1], form light tubes or light strips (bottles) on the optical axis [126, 127], have the property of self-healing after distortion by a small obstacle [128, 129].

The Bessel beams have an orbital angular momentum [130, 131]. A superposition of Bessel beams may have a longitudinal periodicity (analogous to the Talbot effect) [115, 132] or rotate around the optical axis during propagation [133, 134].

The Bessel beams can be generated using digital holograms [126, 127, 13 5], a conical refractive axicon [136, 137], a diffraction vortex axicon [138], diffraction optical elements [133, 134] and spatial light modulators [139]. Interestingly, a simple inclination of the diffraction element (or with an oblique incidence of the illuminating beam on the diffraction vortex axicon) can form Bessel astigmatic beams [140].

The Bessel beams are widely used. They are used for the manipulation of microparticles: for the simultaneous capture of several microparticles along the optical axis [141, 142], for the rotation of one or several particles around the optical axis [143]. Using the Bessel beams, individual cooled atoms can be captured and accelerated [144, 145]. Recently, the open Hankel–Bessel beams [146] can be used for atmospheric probing, since they are resistant to atmospheric turbulence [147]. In [148, 149], the Bessel vector beams were theoretically considered, for which analytical expressions for the OAM density were obtained [130, 131, 150]. We note here that, since the energy of the entire Bessel beam is not limited, the total OAM is also not limited. Therefore, there were no works on analyzing the OAM of the complete Bessel beam before our works. The Bessel modes are also eigenfunctions for circular billiards and correspond to resonant geometric modes that also have an OAM [151].

Recently, the nonparaxial asymmetric Bessel modes [88] and paraxial asymmetric Bessel–Gauss beams [124] were considered. In their cross section, the intensity distribution has the shape of a crescent. In [152], the Bessel asymmetric modes were studied experimentally using a digital micromirror device. In [153], by analogy with [88], asymmetric Chebyshev–Bessel beams were considered.

The superposition of the Bessel beams was considered only for one axial Bessel beams [132–134, 141, 150]. The superposition of laser beams shifted from the optical axis was considered in [151], but these beams are not Bessel beams.

In this section, we consider the superposition of the Bessel beams of the same order shifted from the optical axis (with the same topological charge). A general analytical expression is obtained for the OAM of such a superposition. It is shown that if the weight coefficients of the superposition are real numbers, then the OAM of the whole superposition of the Bessel beams is equal to the OAM of one unshifted Bessel beam. This allows the formation of diffraction-free beams with a different intensity distribution, but with the same OAM. It is shown that the superposition of a set of identical Bessel beams, whose centres are located on a circle of any radius, is equivalent to one Bessel beam from this superposition, located in the centre of the circle. It was also shown that the complex displacement of the Bessel beam leads to a change in the intensity distribution in the beam cross section and a change in its OAM. The superposition of two Bessel beams with a complex shift may not change the OAM, although the intensity distribution will vary. The simulation results are in good agreement with the experimental data.

3.3.1. Fourier spectrum of a shifted Bessel beam

It is known that the complex amplitude of a nonparaxial stationary light field, $E(x,y,z)$ satisfying the Helmholtz equation, can be represented as the angular spectrum of plane waves.

$$E(x,y,z) = \iint_{\mathbb{R}^2} A(\xi,\eta)\exp\left[ik(\xi x+\eta y)+ikz\sqrt{1-\xi^2-\eta^2}\right]d\xi d\eta, \quad (3.50)$$

where k is the wave number of monochromatic light, $A(\xi, \eta)$ is the complex amplitude of the angular spectrum of plane waves. In the polar coordinates (r, φ) the expression (3.50) takes the form:

$$E(r,\phi,z) = \iint_{\mathbb{R}} A(\rho,\theta)\exp\left[ikr\rho\cos(\theta-\phi)+ikz\sqrt{1-\rho}\right]\rho d\rho d\theta, \quad (3.51)$$

where (ρ, θ) are the polar coordinates in the Fourier plane. The amplitude of the angular spectrum of plane waves for a beam shifted by a vector with the Cartesian coordinates (x_0, y_0) has the form:

$$A'(\rho,\theta) = A(\rho,\theta)\exp\left[-ik\rho(x_0\cos\theta+y_0\sin\theta)\right], \quad (3.52)$$

where $A(\rho,\theta)$ is the amplitude of the angular spectrum of plane waves of the initial unshifted beam. The displacement coordinates (x_0, y_0) can be complex numbers.

The form of the angular spectrum of the nth order unshifted Bessel beam is known:

$$A_n(\rho,\theta) = \frac{(-i)^n}{\alpha\lambda}\exp(in\theta)\delta\left(\rho - \frac{\alpha}{k}\right), \qquad (3.53)$$

where $\delta(x)$ is the Dirac delta function, α is the scale parameter of the unshifted Bessel mode:

$$E_n(r,\varphi,z) = \exp\left(in\varphi + iz\sqrt{k^2 - \alpha^2}\right)J_n(\alpha r), \qquad (3.54)$$

where $J_n(x)$ is the Bessel function of the nth order of the first kind. Taking into account (3.52), the amplitude of the angular spectrum of an nth-order shifted Bessel beam is equal to:

$$A'_n(\rho,\theta) = \frac{(-i)^n}{\alpha\lambda}\exp(in\theta)\delta\left(\rho - \frac{\alpha}{k}\right)\exp(-ikx_0\rho\cos\theta - iky_0\rho\sin\theta). \quad (3.55)$$

3.3.2. Connection of the amplitudes of the spectrum of the shifted and unshifted Bessel beams

Let us find the coefficients A_{mn} of the series, which describes the expansion of the amplitude of the spectrum of the nth-order shifted Bessel beam (3.55) in the amplitudes of the spectrum of the unshifted Bessel beams of different orders:

$$A'_n(\rho,\theta) = \delta\left(\rho - \frac{\alpha}{k}\right)\sum_{p=-\infty}^{\infty} A_{pn}\frac{(-i)^p}{\alpha\lambda}\exp(ip\theta). \qquad (3.56)$$

Multiplying both sides of (3.56) by $(\alpha\lambda/\delta\ (\rho - \alpha/k))\exp(-im\ \theta)$ and integrating over θ from 0 to 2π we obtain

$$\int_0^{2\pi}\exp\left[i(n-m)\theta\right]\exp(-i\alpha x_0\cos\theta - i\alpha y_0\sin\theta)d\theta =$$

$$= i^n\sum_{p=-\infty}^{\infty} A_{pn}(-i)^p\ 2\pi\delta_{pm}. \qquad (3.57)$$

The integral on the left side of (3.57) is

$$\int_0^{2\pi}\exp(in\theta + ia\cos\theta + ib\sin\theta)d\theta = 2\pi\left(\frac{ia - b}{\sqrt{a^2 + b^2}}\right)^n J_n\left(\sqrt{a^2 + b^2}\right). \quad (3.58)$$

Taking into account (3.58), the coefficients of the series on the right-hand side of (3.57) are equal to:

$$A_{mn} = \left(\frac{x_0 + iy_0}{\sqrt{x_0^2 + y_0^2}} \right)^{n-m} J_{m-n}\left(\alpha\sqrt{x_0^2 + y_0^2} \right). \tag{3.59}$$

In the particular case when the displacement is real in one coordinate and purely imaginary in the other, and both displacements are equal in modulus ($x_0 = c/\alpha$, $y_0 = ic/\alpha$), the coefficients (3.59) are simplified:

$$A_{mn} = \begin{cases} \dfrac{c^{m-n}}{(m-n)!}, & m \geq n, \\ 0, & m < n. \end{cases} \tag{3.60}$$

Expression (3.60) was obtained from (3.59) using the passage to the limit (for small values of the argument of the Bessel function). In this particular case, the amplitude of the angular spectrum of plane waves of a shifted Bessel beam is represented as a linear combination of the Bessel modes in the form:

$$A_n'(\rho,\theta) = \delta\left(\rho - \frac{\alpha}{k} \right) \sum_{p=0}^{\infty} \frac{c^p}{p!} \frac{(-i)^{n+p}}{\alpha\lambda} \exp\left[i(n+p)\theta \right], \tag{3.61}$$

where the parameter c sets the degree of asymmetry of the nth order Bessel shifted mode The angular spectrum from the placed Bessel beam of the form (3.61) coincides with the expression for the spectrum of the asymmetric Bessel mode [88]. Using (3.55) and (3.58) from (3.50) we obtain an obvious expression for the amplitude of the shifted Bessel beam:

$$E_n'(x,y,z) = \exp\left(iz\sqrt{k^2 - \alpha^2} \right) \left[\frac{(x+x_0) + i(y+y_0)}{\sqrt{(x+x_0)^2 + (y+y_0)^2}} \right]^n \times$$
$$\times J_n\left(\alpha\sqrt{(x+x_0)^2 + (y+y_0)^2} \right). \tag{3.62}$$

3.3.3. Orbital angular momentum of a shifted Bessel beam

The projection on the optical axis z of the orbital angular momentum (OAM) and the total power of the laser beam can be calculated using the expressions:

$$iJ_z = \iint_{\mathbb{R}^2} E^* \frac{\partial E}{\partial \phi} r dr d\phi = \left(\frac{2\pi}{k}\right)^2 \iint_{\mathbb{R}^2} A^* \frac{\partial A}{\partial \theta} \rho d\rho d\theta, \qquad (3.63)$$

$$I = \iint_{\mathbb{R}^2} E^* E r dr d\phi = \left(\frac{2\pi}{k}\right)^2 \iint_{\mathbb{R}^2} A^* A \rho d\rho d\theta, \qquad (3.64)$$

In calculating (3.63) and (3.64) for the shifted Bessel beam, we use (3.55). Then we get the projection of the OAM on the optical axis:

$$J_z = \frac{\lambda}{\alpha} \delta(0) \left[nI_0 (2\alpha r_{0i}) + \frac{\alpha}{r_{0i}} \mathrm{Im}\left(y_0 x_0^*\right) I_1 (2\alpha r_{0i}) \right], \qquad (3.65)$$

where $r_{0i} = [(\mathrm{Im} x_0)^2 + (\mathrm{Im} y_0)^2]^{1/2}$, $I_0(x)$, $I_1(x)$ − are the modified Bessel functions, as well as the power of the whole beam:

$$I = \frac{\delta(0)}{k\alpha} \int_0^{2\pi} \exp\left[2\alpha \left(\mathrm{Im}\, x_0 \cos\theta + \mathrm{Im}\, y_0 \sin\theta \right) \right] d\theta =$$
$$= \frac{\lambda}{\alpha} \delta(0) I_0 (2\alpha r_{0i}), \qquad (3.66)$$

where $\delta(0)$ is the Dirac delta function at zero. From (3.65) and (3.66) it follows that the projection of the OAM on the optical axis and the power of the shifted Bessel beam are unbounded, but their ratio has a well-defined value:

$$\frac{J_z}{I} = n + \frac{\alpha}{r_{0i}} \mathrm{Im}\left(x_0^* y_0\right) \frac{I_1 (2\alpha r_{0i})}{I_0 (2\alpha r_{0i})}. \qquad (3.67)$$

From (3.67) it follows that if both coordinates of the displacement vector (x_0, y_0) are real or both purely imaginary, then the normalized OAM (3.67) does not differ from the OAM for the unshifted Bessel beam:

$$\frac{J_z}{I} = n. \qquad (3.68)$$

The normalized OAM of the shifted Bessel beam (3.67) will differ from the OAM of the unshifted Bessel beam (3.68) only if the displacement along one coordinate is real, and the other is purely imaginary. For example, let $x_0 = b/\alpha$, $y_0 = ic/\alpha$, then the OAM (3.67) takes the form:

$$\frac{J_z}{I} = n + b\frac{I_1(2|c|)}{I_0(2|c|)}. \tag{3.69}$$

From (3.69) it follows that the OAM will increase compared to (3.68) if $b > 0$, and decrease if $b < 0$. Moreover, the magnitude of the imaginary displacement c determines the change in beam shape: for small values of $c < 1$ the amplitude from the shifted beam (3.62) will have the form of an ellipse, with $a > 1$, the amplitude of the beam (3.62) will have the form of the crescent, and at $c \gg 1$ the beam (3.62) will have the form an astigmatic Gaussian beam [88]. The centre of the Bessel beam (3.54) is shifted along the x axis by the value $\Delta x = (b-c)/\alpha$.

3.3.4. Orbital angular momentum of superposition of Bessel's shifted beams

Consider the superposition P of the nth order shifted Bessel beams of order (3.62). The amplitude of the angular spectrum of plane waves for such a superposition will be:

$$A(\rho,\theta) = \sum_{p=0}^{P-1} C_p A_{pn}(\rho,\theta), \tag{3.70}$$

where

$$A_{pn}(\rho,\theta) = \frac{(-i)^n}{\alpha\lambda}\exp(in\theta)\delta\left(\rho - \frac{\alpha}{k}\right) \times$$
$$\times\exp\left(-ikx_p\rho\cos\theta - iky_p\rho\sin\theta\right) \tag{3.71}$$

is the amplitude of the angular spectrum of the p-th beam in a superposition shifted by a complex vector with coordinates (x_p, y_p). Using (3.63) and (3.64), we obtain the normalized OAM of the superposition (3.70).

$$\frac{J_z}{I} = n - i\alpha \frac{\displaystyle\sum_{p=0}^{P-1}\sum_{q=0}^{P-1} C_p^* C_q \frac{x_p^* y_q - x_q y_p^*}{R_{pq}} J_1(\alpha R_{pq})}{\displaystyle\sum_{p=0}^{P-1}\sum_{q=0}^{P-1} C_p^* C_q J_0(\alpha R_{pq})}, \qquad (3.72)$$

where $J_0(x)$, $J_1(x)$ are Bessel functions of zero and first orders,

$$R_{pq} = \sqrt{(x_p^* - x_q)^2 + (y_p^* - y_q)^2},$$
$$R_{pp} = 2i\sqrt{(\mathrm{Im}\, x_p)^2 + (\mathrm{Im}\, y_p)^2}. \qquad (3.73)$$

Although there is a factor $i\alpha$ before the fraction in (3.72), the whole expression is real. This follows from the fact that: 1) when $p = q$ in the numerator, the quantity $|C_p|^2$ is real, R_{pq} and $J_1(\alpha R_{pq})$ are purely imaginary, and the difference $x_p^* y_p - x_p y_p^*$ is also purely imaginary, since it is the difference of two complex conjugate numbers; 2) for any p and q unequal to each other, $R_{pq} = R_{qp}^*$, and the terms with the numbers (p, q) and (q, p) are also the difference of two complex-conjugate numbers.

It can be shown that if all Bessel beams in superposition (3.70) are shifted by a real vector (x_p, y_p) and all coefficients C_p are also real, then the numerator of the fraction in (3.72) will be equal to zero, OAM of one unshifted n-th order Bessel beam (3.68). This is the main result of this section. It allows the formation of a wide variety of nonparaxial laser beams, which will have different intensity distributions in the beam section, but will have one and the same OAM (3.68), and will be distributed without diffraction. Below the modelling section presents examples of such beams.

Interesting special cases follow from (3.72). If $P = 2$, $x_0 = c/\alpha$, $y_0 = ic/\alpha$, $x_1 = -c/\alpha$, $y_1 = ic/\alpha$, then $R_{00} = R_{11} = 2ic/\alpha$, $R_{01} = R_{10} = 0$, and for a normalized OAM we obtain a simple form (coefficients in (3.70) arbitrary complex numbers C_0, C_1):

$$\frac{J_z}{I} = n + \frac{c\left(|C_0|^2 - |C_1|^2\right) I_1(2|c|)}{\left(|C_0|^2 + |C_1|^2\right) I_0(2|c|) + 2\,\mathrm{Re}\{C_0^* C_1\}}. \qquad (3.74)$$

From (3.74) it follows that with the addition of two Bessel beams of the nth order, the displacement of which, although complex (purely imaginary in one coordinate), but consistent with each other, with coefficients of equal modulus $|C_0| = |C_1|$ the normalized OAM (3.74) will be equal to the OAM of one unshifted Bessel beam of the nth order (3.68). That is, with equal coefficients $|C_0| = |C_1|$ it is possible to change the intensity distribution in the cross section of the superposition of two shifted Bessel beams (since the shape of the Bessel beam changes as the value of c changes), although their total OAM will not change.

3.3.5. *Superposition of three shifted Bessel beams*

Consider three Bessel beams of order n, shifted so that their centres are in the corners of an equilateral triangle. That is, $P = 3$, $R_{01} = R_{02} = R_{12}$, the weight coefficients of the superposition (3.70) C_0, C_1, C_2 are arbitrary complex numbers, and let the coordinates of the complex displacement vector be given by the expression:

$$\begin{cases} x_p = R_0 \cos\left(\dfrac{2\pi p}{3}\right) + \dfrac{c}{\alpha}\exp\left(-i\gamma - i\dfrac{2\pi p}{3}\right), \\[2mm] y_p = R_0 \sin\left(\dfrac{2\pi p}{3}\right) + i\dfrac{c}{\alpha}\exp\left(-i\gamma - i\dfrac{2\pi p}{3}\right), \end{cases} \tag{3.75}$$

where R_0 is the radius of the circle on which the singularity centres of the shifted Bessel beams are located, with c setting the asymmetry of the shifted Bessel beam, γ is the angle of rotation of the asymmetric Bessel shifted beam. Then OAM will be equal to the expression:

$$\frac{J_z}{I} = n + \frac{\pm D_1 \xi I_1(2c) + \text{Im}\left\{D_2\left(\xi \mp ic\sqrt{3}\right)J_1\left(\sqrt{3}\xi \pm ic\right)\right\}}{D_1 I_0(2c) + 2\text{Re}\left\{D_2 J_0\left(\sqrt{3}\xi \pm ic\right)\right\}}, \tag{3.76}$$

where the upper sign is taken for $\gamma = 0$ and the lower sign is taken for $\gamma = \pi$ (for other values of γ, the expression for the normalized OAM will be more cumbersome),

$$\begin{cases} D_1 = |C_0|^2 + |C_1|^2 + |C_2|^2, \\[1mm] D_2 = C_0^* C_1 + C_1^* C_2 + C_2^* C_0, \end{cases} \tag{3.77}$$

wherein $\xi = \alpha R_0 \pm c$.

In the particular case, when $\gamma = \pi$, $c = \alpha R_0$, the parameter ξ is zero, and instead of (3.76) we get:

$$\frac{J_z}{I} = n + \frac{c\sqrt{3}I_1(c)\operatorname{Im}\{D_2\}}{D_1 I_0(2c) + 2I_0(c)\operatorname{Re}\{D_2\}}. \tag{3.78}$$

From (3.78) it follows that if the coefficients C_0, C_1, C_2 are real, then the factor $\operatorname{Im}\{D_2\}$ in (3.78) is zero and the OAM of the superposition of the nth-order three Bessel beams is equal to the OAM of one unshifted beam Bessel (3.68). For example, Fig. 3.11 shows the intensity and phase of the superposition of the three shifted Bessel beams with a topological charge $n = 3$. The OAM of the superposition is equal to $J_z/I = 3$

Figure 3.12 shows the coded phase (Fig. 3.12 *a*) of a superposition of the three shifted Bessel beams with a topological charge $n = 3$ (Fig. 3.11 *b*). This phase was fed to a spatial light modulator SLM PLUTO-VIS (resolution 1920 × 1080 pixels, pixel size 8 μm).

Figures 3.12 *b–d* show the intensity distributions formed by the SLM upon reflection of a linearly polarized plane wave with a wavelength of 633 nm at different distances. From Fig. 3.12 it is clear that the beam retains its structure during propagation, and the intensity distribution is consistent with the calculated one (Fig. 3.11 *a*). The intensity was recorded using an MDCE-5A CMOS camera (1/2", resolution 1280 × 1024 pixels).

Figure 3.13 shows the distribution of intensity and superposition of the three phases of the shifted Bessel beams of the same unit weight coefficients $\mathbf{C} = [1, 1, 1]$, but for other parameter values: $n = 5$, $R_0 = 8\lambda$, $c = 3$. The diffraction of such a beam is quite different – instead of a light triangle three light spots are formed. For such a beam $c \neq \alpha R_0$, therefore, the OAM can not be determined

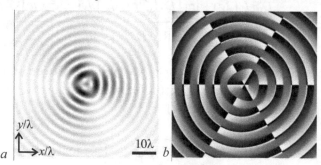

Fig. 3.11. The intensity (*a*) (black colour – maximum, white – zero) and the phase (*b*) (black – π, white – $-\pi$) superposition of three shifted Bessel modes with parameters: $n = 3$, $R_0 = 4\lambda$, $\alpha = 1/\lambda$, $c = 4$, $\gamma = \pi$, vector of weight coefficients $\mathbf{C} = [1, 1, 1]$. Frame size $2R = 60\lambda$,

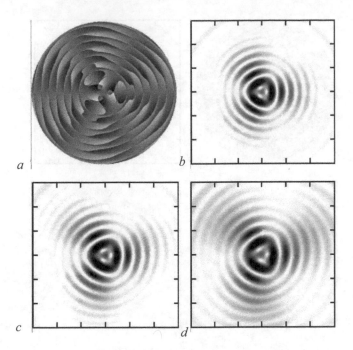

Fig. 3.12. The coded phase (Fig. 3.11 *b*, negative) for the formation of a Bessel beam as an equilateral triangle contour (*a*) and experimentally formed intensity distributions (negatives) at various distances from the plane $z = 0$: (*b*) 0 mm; (*c*) 200 mm; (*d*) 400 mm. The grid spacing is 0.5 mm.

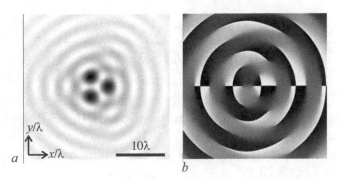

Fig. 3.13. Intensity (*a*) (black – maximum, white colour – zero) and phase (*b*) (black colour – π, white colour – $-\pi$) superposition of three Bessel beams with parameters $n = 5$, $R_0 = 8\lambda$, $\alpha = 1/\lambda$, $c = 3$, $\gamma = \pi$, weight vector **C** = [1, 1, 1]. Frame size $2R = 30\lambda$.

using formula (3.78). According to the formula (3.76), the OAM of this beam is fractional and is equal to

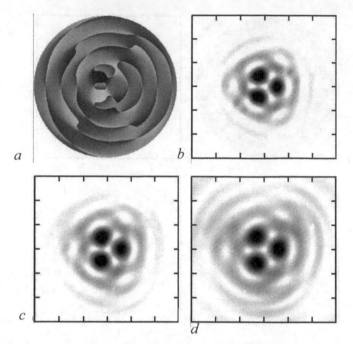

Fig. 3.14. The coded phase for the formation of a Bessel beam in the form of three light spots (Fig. 3.13 *b*, negative) (*a*) and experimentally formed intensity distributions (negatives) at different distances from the plane $z = 0$: (*b*) 0 mm; (*c*) 200 mm; (*d*) 400 mm. The grid spacing is 0.5 mm.

$$\frac{J_z}{I} = 5 + \frac{\mathrm{Im}\left\{\left(5 + i3\sqrt{3}\right)J_1\left(5\sqrt{3} - 3i\right)\right\} - 5I_1(6)}{I_0(6) + 2\mathrm{Re}\left\{J_0\left(5\sqrt{3} - 3i\right)\right\}} \approx 0.62. \qquad (3.79)$$

Figure 3.14 shows the phase (*a*) and intensity distribution (*b–d*) formed by the SLM, for the superposition of three Bessel beams with parameters $n = 5$, $R_0 = 8\lambda$, $\alpha = 1/\lambda$, $c = 3$, $\gamma = \pi$. The figure shows that the diffraction patterns are consistent with the calculated intensity in Fig. 3.13 *a*.

3.3.6. Superposition of Bessel beams located at the corners of a regular polygon

As in the previous case, we consider the superposition of P shifted Bessel beams of n-th order, singularity centres are located at the vertices of a regular polygon (similar to (3.75)):

$$
\begin{cases}
x_p = R_0 \cos\left(\dfrac{2\pi p}{P}\right) + \dfrac{c}{\alpha}\exp\left(-i\gamma - i\dfrac{2\pi p}{P}\right), \\[4mm]
y_p = R_0 \sin\left(\dfrac{2\pi p}{P}\right) + i\dfrac{c}{\alpha}\exp\left(-i\gamma - i\dfrac{2\pi p}{P}\right),
\end{cases}
\tag{3.80}
$$

where $p = 0,\ldots, P - 1$.

For definiteness, let $\gamma = \pi$ and $c = \alpha R_0$. In this case, instead of (3.80) will be:

$$
\begin{cases}
x_p = iR_0 \sin\left(\dfrac{2\pi p}{P}\right), \\[4mm]
y_p = -iR_0 \cos\left(\dfrac{2\pi p}{P}\right).
\end{cases}
\tag{3.81}
$$

The general formula for OAM (3.72) will look like:

$$
\frac{J_z}{I} = n + \alpha \times
$$

$$
\times \frac{2R_0 \sum\limits_{p=1}^{P-1}\sum\limits_{q=0}^{p-1} \operatorname{Im}\left\{C_p^* C_q\right\} \sin\left[\dfrac{\pi(p-q)}{P}\right] I_1\left\{2\alpha R_0 \cos\left[\dfrac{\pi(p-q)}{P}\right]\right\}}{\sum\limits_{p=0}^{P-1}\left|C_p\right|^2 I_0\left(2\alpha r_{pi}\right) + 2\sum\limits_{p=1}^{P-1}\sum\limits_{q=0}^{p-1} \operatorname{Re}\left\{C_p^* C_q J_0\left(\alpha R_{pq}\right)\right\}}.
\tag{3.82}
$$

If all the coefficients C_p are real, then the numerator in (3.82) is zero and the OAM of such a superposition of shifted Bessel

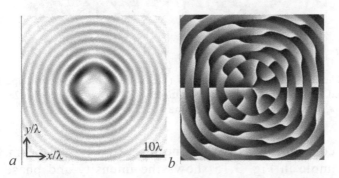

Fig. 3.15. Intensity (*a*) (black colour – maximum, white colour – zero) and phase (*b*) (black colour – π, white colour – $-\pi$) are superpositions of four ($P = 4$) shifted Bessel beams with parameters $n = 7$, $R_0 = 6\lambda$, $\alpha = 1/\lambda$, $c = 6$, $\gamma = \pi$, $\mathbf{C} = [1, 1, 1, 1]$. Frame size $2R = 60\lambda$.

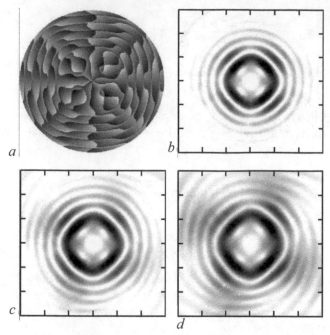

Fig. 3.16. Coded phase (Fig. 3.15 *b*, negative) for the formation of a Bessel beam in the form of a square contour (*a*) and experimentally formed intensity distributions at different distances from the plane $z = 0$: (*b*) 0 mm; (*c*) 200 mm; (*d*) 400 mm. The grid spacing is 0.5 mm.

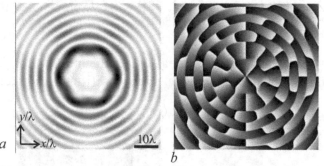

Fig. 3.17. The intensity (*a*) and phase (*b*) (black – π, white – $-\pi$) of superposition of six shifted Bessel beams with the parameters: $P = 6$, $n = 10$, $R_0 = 12\lambda$, $\alpha = 1/\lambda$, $c = 12$, $\gamma = \pi$, $\mathbf{C} = [1, 1, 1, 1, 1, 1]$. Frame size $2R = 60\lambda$.

beams is equal to the OAM of one unshifted Bessel beam (3.68). As an example in Fig. 3.15 shows the intensity and phase of the superposition of the four shifted Bessel beams with topological charges $n = 7$. The normalized OAM of this superposition is $J_z/I = 7$. From Fig. 3.15 *b* it is clear that inside the main bright ring, which

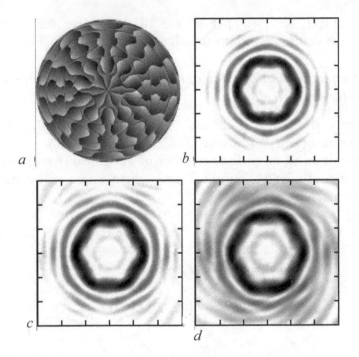

Fig. 3.18. Coded phase (Fig. 3.17 *a,* negative) for the formation of a Bessel beam in the form of a hexagon contour (*a*) and experimentally formed intensity distributions at different distances from the plane $z = 0$: (*b*) 0 mm; (*c*) 200 mm; (*d*) 400 mm. The grid spacing is 0.5 mm.

looks like a square, there are 7 optical vortices with a topological charge +1.

Figure 3.16 shows the coded phase (*a*) and intensity distribution (*b–d*) formed by the SLM, for the superposition of the four shifted Bessel beams (Fig. 3.15) with the topological charge $n = 7$, $R_0 = 6\lambda$, $\alpha = 1/\lambda$, $c = 6$, $\gamma = \pi$, $\mathbf{C} = [1, 1, 1, 1]$. From Fig. 3.16 it can be seen that the diffraction patterns are consistent with the calculated intensity in Fig. 3.15 *a.*

Figure 3.17 shows another example: the intensity (*a*) and phase (*b*) of a superposition of six ($P = 6$) shifted Bessel beams with the same topological charge $n = 10$, whose singularity centres are located at the vertices of a regular hexagon. The normalized OAM of such a superposition is equal to $J_z/I = 10$.

Figure 3.18 shows the coded phase (*a*) and the intensity distribution (*b–d*) formed by SLM at different distances for a superposition of six ($P = 6$) shifted Bessel beams (Fig. 3.17) with the same topological charge $n = 10$, $R_0 = 12\,\lambda$, $\alpha = 1/\lambda$, $c = 12$, γ

$= \pi$, $\mathbf{C} = [1, 1, 1, 1, 1, 1]$. Figure 3.18 shows that the diffraction patterns are consistent with the calculated intensity in Fig. 3.17 *a*.

3.3.7. Superposition of a large number of Bessel beams whose centres lie on a circle

This section shows that if at each point of a circle of radius R_0 there is the centre of an nth order shifted Bessel beam, and we consider a superposition of an infinite number of such beams with identical weighting coefficients, then the entire superposition will form an nth order ordinary Bessel beam.

Using the Bessel function as a series, the unshifted Bessel beam (3.54) in the initial plane ($z = 0$) can also be represented as a series:

$$E(x,y,z=0) = \exp(in\varphi) J_n(\alpha r) =$$

$$= \sum_{p=0}^{\infty} \frac{(-1)^p}{p!(n+p)!}\left(\frac{\alpha}{2}\right)^{n+2p}(x-iy)^p(x+iy)^{n+p}. \tag{3.83}$$

Consider a continuous superposition of shifted Bessel beams (3.83) whose centres are located on a circle of radius R_0:

$$E(x,y,z=0) =$$

$$= \sum_{p=0}^{\infty} \frac{(-1)^p}{p!(n+p)!}\left(\frac{\alpha}{2}\right)^{n+2p} \int_0^{2\pi} \left[(x-R_0\cos\theta)-i(y-R_0\sin\theta)\right]^p \times \tag{3.84}$$

$$\times \left[(x-R_0\cos\theta)+i(y-R_0\sin\theta)\right]^{n+p} d\theta.$$

In the polar coordinates (3.84) will look like:

$$E(r,\varphi,z=0) =$$

$$= \sum_{p=0}^{\infty} \frac{(-1)^p}{p!(n+p)!}\left(\frac{\alpha}{2}\right)^{n+2p} \int_0^{2\pi} \left[r\exp(-i\varphi)-R_0\exp(-i\theta)\right]^p \times \tag{3.85}$$

$$\times \left[r\exp(i\varphi)-R_0\exp(i\theta)\right]^{n+p} d\theta.$$

Imagine both factors under the integral in (3.85) as a binomial decomposition:

$$E(r,\varphi,z=0)=$$

$$=\sum_{p=0}^{\infty}\frac{(-1)^p}{p!(n+p)!}\left(\frac{\alpha}{2}\right)^{n+2p}\sum_{m=0}^{p}\sum_{k=0}^{n+p}\binom{p}{m}\binom{n+p}{k}(-R_0)^{m+k}r^{2p-m+n-k}\times$$

$$\exp\left[i(n+p-k)\varphi-i(p-m)\varphi\right]\int_{0}^{2\pi}\exp\left[i(k-m)\theta\right]d\theta. \tag{3.86}$$

Since the integral over θ is not equal to zero only when $k = m$, the sum over k disappears, and instead of (3.86) we get:

$$E(r,\varphi,z=0)=2\pi\exp(in\varphi)\sum_{p=0}^{\infty}\frac{(-1)^p}{p!(n+p)!}\left(\frac{\alpha r}{2}\right)\times$$

$$\times \; {}^{n+2p}\sum_{m=0}^{p}\binom{p}{m}\binom{n+p}{m}\left(\frac{R_0}{r}\right)^{2m}. \tag{3.87}$$

Changing the order of summation in (3.87), we get:

$$E(r,\varphi,z=0)=2\pi\exp(in\varphi)=$$

$$=\left[\sum_{m=0}^{\infty}\left(\frac{R_0}{r}\right)^{2m}\sum_{p=m}^{\infty}\binom{p}{m}\binom{n+p}{m}\frac{(-1)^p}{p!(n+p)!}\left(\frac{\alpha r}{2}\right)^{n+2p}\right]=$$

$$=2\pi\exp(in\varphi)\sum_{m=0}^{\infty}\left(\frac{R_0}{r}\right)^{2m}\sum_{p=m}^{\infty}\frac{(-1)^p}{(m!)^2(p-m)!(n+p-m)!}\left(\frac{\alpha r}{2}\right)^{n+2p}. \tag{3.88}$$

Replace $p-m$ by p, and rearrange the factors in two rows so that it becomes clear that these series are equal to the Bessel functions:

$$E(r,\varphi,z=0)=2\pi\exp(in\varphi)=$$

$$=\left[\sum_{m=0}^{\infty}\frac{(R_0/r)^{2m}}{(m!)^2}\sum_{p=0}^{\infty}\frac{(-1)^{p+m}}{p!(n+p)!}\left(\frac{\alpha r}{2}\right)^{n+2p+2m}\right]=$$

$$=2\pi\exp(in\varphi)\left[\sum_{m=0}^{\infty}\frac{(-1)^m}{(m!)^2}\left(\frac{\alpha R_0}{2}\right)^{2m}\right]\left[\sum_{p=0}^{\infty}\frac{(-1)^p}{p!(n+p)!}\left(\frac{\alpha r}{2}\right)^{n+2p}\right]= \tag{3.89}$$

$$=2\pi J_0(\alpha R_0)J_n(\alpha r)\exp(in\varphi).$$

Fig. 3.19. Superposition intensity P of Bessel beams with $P = 5$ (*a*), 8 (*b*), 10 (*c*), 20 (*d*), 40 (*e*), 60 (*f*) (black colour – maximum, white colour – zero).

Amplitude (3.89) differs from the amplitude of the Bessel beam normal factor $2\pi J_0 (\alpha R_0)$. Figure 3.19 shows the superposition of intensity for 5, 8, 10, 20, 40 and 60 shifted Bessel beams with the topological charge $n = 7$, the centres of which lie on a circle of radius $R_0 = 100\lambda$. Calculation parameters for all in Figs. 3.19 *a–f* are identical: $n = 7$, $R_0 = 100\lambda$, $\alpha = 1/\lambda$. Frame size $2R = 240\lambda$.

Figure 3.19 shows that the unshifted Bessel mode with the amplitude (39) is almost already formed at $P = 60$. The normalized OAMs for all superpositions in Figs. 3.19 *a–f* are the same and equal to $J_z/I = 7$. From (3.89) it can be seen that for $\alpha R_0 = \gamma_{0,s}$, ($\gamma_{0,s}$ is the s-th root of the zero-order Bessel function) the complex amplitude is zero (i.e., all Bessel beams experience destructive interference). Numerical simulation confirms this: for superposition of $P = 360$ beams with $R_0 = 100\lambda$ and $R_0 = \gamma_{0.32}\lambda = 99.74682\lambda$ the obtained images similar to Fig. 3.1 *f*, however, the intensity in the second case is 3×10^{12} times less.

In this section, an analytical expression is obtained for calculating the normalized OAM for the superposition of Bessel beams shifted from the optical axis with the same topological optical charge (expression (3.72)). It is shown that if the weight coefficients of the superposition are real, then the OAM of the entire superposition of the Bessel beams is equal to the OAM of one unshifted Bessel beam. This allows the formation of diffraction-free beams with a different intensity distribution, but with the same OAM. It is shown that the superposition of a set of identical Bessel beams, the centres of which are located on a circle of any radius, is equivalent to one Bessel beam from this superposition located at the centre of the circle (expression (3.89)). It was also shown that the complex displacement of the Bessel beam leads to a change in the intensity distribution in the beam cross section and a change in its OAM (expressions (3.67), (3.69)). The superposition of two Bessel beams with a complex displacement may not change the OAM, although the intensity distribution will vary (expression (3.74)). The experiment agrees well. with theory (Figs. 3.12, 3.14, 3.16, 3.18).

3.4. Asymmetric Bessel modes of other types

The nonparaxial asymmetric Bessel modes (aB-modes) and paraxial asymmetric Bessel–Gauss beams were considered in [87, 88, 155, 124]. In the cross section of these laser beams, the intensity distribution has the form of a crescent. In [152], the aB-modes were studied experimentally using a digital micromirror device. In [153], by analogy with [88] (introducing a complex displacement of the Bessel mode), asymmetric Chebyshev–Bessel beams were considered. In [156], it is proposed to use aB-beams as acoustic vortex beams. In [157], vectorless diffraction beams with a fractional orbital angular momentum, including asymmetric (Mathieu and Weber) beams, analogous to [88], were studied.

In this section, it is shown that aB-beams [88] are generated by the conventional symmetric Bessel and Bessel–Gauss modes and by moving the argument of the Bessel function in the complex plane. The asymmetric Bessel beams of the second type, differing from

[88] by the type of complex displacement, are also considered. The superposition of the aB-beams of the first (I) and second (II) type is also considered, and it is shown that, although for the aB-beams of both types the OAM is fractional and depends on the asymmetry parameter c, for the sum and difference of aB-beams I and II whole and equal to the topological charge of the Bessel mode n for any value of c. Hence, it is possible to form nonparaxial modes with different transverse intensity distributions (symmetric or asymmetrical) with the same whole OAM.

3.4.1. Asymmetric Bessel modes of type II

The aB-modes considered earlier have the form [88]:

$$E_1(r,\varphi,c) = \left[\frac{\alpha r}{\alpha r - 2c\exp(i\varphi)}\right]^{n/2}$$
$$J_n\left\{\sqrt{\alpha r(\alpha r - 2c\exp(i\varphi))}\right\}\exp(in\varphi), \qquad (3.90)$$

where $J_n(x)$ is the nth order Bessel function of the first kind, (r, φ) are polar coordinates, α is a scale factor, c is a dimensionless coefficient, in general, complex. It can be shown that mode (3.90) is generated by the usual Bessel mode

$$E_1(r,\varphi,c=0) = J_n(\alpha r)\exp(in\varphi) \qquad (3.91)$$

when the argument is shifted to the complex area. Indeed, the field (3.90) in the Cartesian coordinates has the form:

$$E_1(x,y,c) = \left(\frac{x+iy}{x-iy-2c/\alpha}\right)^{n/2} J_n\left\{\alpha\sqrt{(x+iy)(x-iy-2c/\alpha)}\right\}. \qquad (3.92)$$

The complex amplitude (3.92) can be reduced to (3.91) by replacing the variables:

$$\begin{cases} y = y' + ic/\alpha, \\ x = x' + c/\alpha. \end{cases} \qquad (3.93)$$

In (3.93), the parameter c will be considered for definiteness a positive real number. The displacement of the mode (3.91) along the horizontal axis (x) is accompanied by its modification, since along the vertical axis (y) the displacement in (3.93) is purely imaginary. This change in the transverse structure of the beam intensity leads to the fact that instead of a light ring, the mode has the form of a weak ellipse ($c < 1$), a growing crescent ($c > 1$) or astigmatic Gaussian beam ($c \gg 1$). From (3.93) it can be seen that in absolute value and in sign the displacements along the x and y axes are the same. But, generally speaking, this is not a mandatory requirement: for the mode (3.91) to remain a mode of displacement along different Cartesian axes, it can be different in magnitude and in sign. But for further considerations, in order to obtain two different modes using the superposition of new Bessel modes, we confine ourselves to two beam modifications (3.90). Consider a beam with a negative value of the parameter $c' = -c(c>0)$. Then instead of (3.90) we get the complex amplitude of the aB-beam, which has a decreasing crescent in the cross-section of intensity:

$$
E_1(r,\varphi,-c) = \left[\frac{\alpha r}{\alpha r + 2c \exp(i\varphi)} \right]^{n/2} =
$$
$$
= J_n\left\{ \sqrt{\alpha r(\alpha r + 2c \exp(i\varphi))} \right\} \exp(in\varphi).
$$

(3.94)

Consider another modification of the Bessel beam (3.90), when the displacements along different Cartesian axes are equal in magnitude, but different in sign:

$$
\begin{cases} y = y' + ic/\alpha, \\ x = x' - c/\alpha, \end{cases}
$$

(3.95)

then we get an aB-beam of type II, which also in the cross-section of intensity gives a growing crescent, but is shifted in relation to the the crescent of the mode (3.90) for the same values of $c > 0$:

$$
E_2(x,y,c) = \left(\frac{x+iy+2c/\alpha}{x-iy} \right)^{n/2} J_n\left\{ \alpha\sqrt{(x-iy)(x+iy+2c/\alpha)} \right\}.
$$

(3.96)

In the polar coordinates, a Bessel type II mode (3.96) will look like:

$$E_2\left(r,\varphi,c\right)=\left[\frac{\alpha r+2c\exp\left(-i\varphi\right)}{\alpha r}\right]^{n/2}J_n\times$$
$$\times\left\{\sqrt{\alpha r\left(\alpha r+2c\exp\left(-i\varphi\right)\right)}\right\}\exp\left(in\varphi\right). \tag{3.97}$$

A type II mirror mode for the mode (3.97) is obtained similarly to mode (3.94) by changing the parameter from c to $-c$:

$$E_2\left(r,\phi,-c\right)=\left[\frac{\alpha r-2c\exp\left(-i\phi\right)}{\alpha r}\right]^{n/2}\times$$
$$\times J_n\left\{\sqrt{\alpha r\left(\alpha r-2c\exp\left(-i\phi\right)\right)}\right\}\exp\left(in\phi\right). \tag{3.98}$$

The asymmetric Bessel mode (3.98) has an intensity in the cross section as a decreasing half-crescent, the same as the mode (3.94), but shifted compared to the initial mode (3.91) by a different distance with the same value of c.

The orbital angular moments of the modes of the first (3.90), (3.94) and the second type (3.97), (3.98) are different:

$$\frac{J_z}{I}=n\pm\frac{cI_1\left(2c\right)}{I_0\left(2c\right)}, \tag{3.99}$$

where $I_n(x)$ is the modified Bessel functions,

$$I = \iint|E(r,\phi)|^2rdrd\phi$$

is the power of the beam. The expression (3.99) for the aB mode I was obtained in [88]. The OAM for aB-mode II is obtained similarly. In (3.99), the upper sign is chosen for beams of the first type (3.90), (3.94), and the lower sign is chosen for beams of the second type (3.97), (3.98).

When propagating in space, the complex amplitudes of the aB-beams (3.90), (3.94) and (3.97), (3.98) are multiplied by the same factor $\exp\left(ikz\sqrt{1-\alpha^2}\right)$.

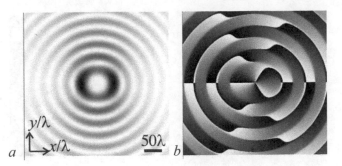

Fig. 3.20. Intensity (a) and phase (b) of the sum of two asymmetric and mutually mirror Bessel modes (3.100).

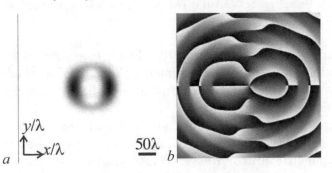

Fig. 3.21. Intensity (*a*) and phase (*b*) of the sum of two asymmetric and mutually mirror Bessel–Gauss beams, with the same parameters as in Fig. 3.20

To obtain the corresponding asymmetric Bessel–Gauss beams of the first and second types, the complex amplitudes (3.90), (3.94) and (3.97), (3.98) must be multiplied by the Gaussian exponent, where w is the Gaussian beam radius.

3.3.2. Superposition of asymmetric Bessel modes

Although the laser beams $E_{1,2}(r,\varphi,\pm c)$ in the cross section of intensity give a crescent, superpositions of such beams can form various Bessel modes, both symmetric and asymmetric. Let us consider several examples of such superpositions of modes $E_{1,2}(r,\varphi,\pm c)$.

Figure 3.20 shows the intensity (*a*) and phase (*b*) of the sum of two mirror asymmetric modes of Bessel type 1:

$$E_{+}(r,\varphi,z;c) = E_{1}(r,\varphi,z;c) + E_{1}(r,\varphi,z;-c). \tag{3.100}$$

Calculating parameters: wavelength λ = 532 nm, the topological charge n = 3, the asymmetry parameter c = 1, scale factor α = 1/

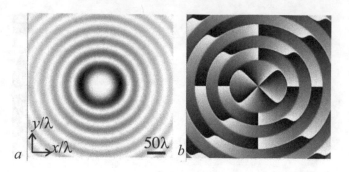

Fig. 3.22. Intensity (*a*) and phase (*b*) of the difference between two mutually mirror asymmetric Bessel modes (3.101). The calculation parameters are the same as in Fig. 3.20.

(10λ), frame resolution $R = 200\lambda$. From Fig. 3.20 it is seen that the resulting mode has the symmetry about the Cartesian axes, the intensity distribution has the form of an ellipse, with a small non-uniformity of intensity.

Figure 3.21 shows the intensity (*a*) and phase (*b*) of the sum (3.100) multiplied by the Gaussian function. It can be seen that the two asymmetric mirror Bessel–Gauss beams with the waist radius $w = 50\lambda$ at $c = 3$ form a symmetrical elliptical beam with almost no side lobes, similar to the letter O.

Interestingly, the difference between the two mirror asymmetric modes of Bessel type 1

$$E_- (r,\varphi,z;c) = E_1 (r,\varphi,z;c) - E_1 (r,\varphi,z;-c) \qquad (3.101)$$

forms the intensity (Fig. 3.22 *a*) and phase (Fig. 3.22 *b*) almost with axial symmetry (weak ellipse), and with an increase by one of the topological charge of the optical vortex on the axis. The parameters are the same as in Fig. 3.20. Figure 3.22 *b* shows that in the centre of a vortex pattern formed on the optical axis with the topological charge $n = 4$, although in both modes (3.101) the topological charge is equal to $n = 3$. In the sum of the modes (3.100) is also formed on the optical vortex axis topological charge $n = 3$ (Fig. 3.20 *b* and Fig. 3.21 *b*).

An increase in the topological charge by one when subtracting two mirror modes (3.90) and (3.94) can be proved. Indeed, if we consider the amplitude of the sum (3.100) and the difference (3.101)

$$E_{\pm}(r,\varphi,z) = (\alpha r)^{n/2} \exp(in\varphi) \times$$

$$\times \left(\frac{J_n\left\{\sqrt{\alpha r\left[\alpha r - 2c\exp(i\varphi)\right]}\right\}}{\left[\alpha r - 2c\exp(i\varphi)\right]^{n/2}} \pm \frac{J_n\left\{\sqrt{\alpha r\left[\alpha r + 2c\exp(i\varphi)\right]}\right\}}{\left[\alpha r + 2c\exp(i\varphi)\right]^{n/2}} \right), \quad (3.102)$$

in a neighbourhood of the origin, then, using an approximate expression for the Bessel function,

$$J_n(z) \approx \frac{1}{n!}\left(\frac{z^2}{4}\right)^{n/2}\left[1 + \frac{z^2}{4(n+1)}\right], \quad (3.103)$$

we will receive:

$$E_{\pm}\left(r \ll \frac{1}{\alpha}, \varphi, z\right) = \frac{1}{n!}(\alpha r)^{n/2}\left(\frac{\alpha r}{4}\right)^{n/2} \exp(in\varphi) \times$$

$$\times \left\{\left[1 + \frac{\alpha r\left[\alpha r - 2c\exp(i\varphi)\right]}{4(n+1)}\right] \pm \left[1 + \frac{\alpha r\left[\alpha r + 2c\exp(i\varphi)\right]}{4(n+1)}\right]\right\}, \quad (3.104)$$

i.e.

$$E_{+} \approx \frac{2}{n!}\left(\frac{\alpha r}{2}\right)^n\left[1 + \frac{(\alpha r)^2}{4(n+1)}\right]\exp(in\varphi),$$

$$E_{-} \approx \frac{-2c}{(n+1)!}\left(\frac{\alpha r}{2}\right)^{n+1} \exp\left[i(n+1)\varphi\right]. \quad (3.105)$$

From (3.105) it can be seen that when adding two modes (3.90) and (3.94) with topological charge n, the vortex E_{+} with topological charge n is obtained near the optical axis, and with subtraction, optical vortex E_{-} with topological charge $n + 1$.

We now consider the sum and difference of two identical modes, but of different types: the first (3.90) and the second (3.97):

$$E_{1+2}(r,\varphi,z) = E_1(r,\varphi,z;c) + E_2(r,\varphi,z;c). \quad (3.106)$$

For small values of the asymmetry parameter ($c = 1$), the sum (3.106) gives almost the same growing crescent (Fig. 3.23 *a*) as each

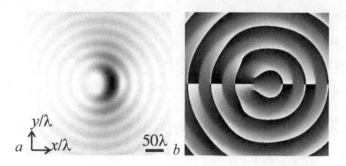

Fig. 3.23. Intensity (*a*) and phase (*b*) of the sum of two identical asymmetric Bessel modes of the first and second types (3.106). The calculation parameters are the same as in Fig. 3.20.

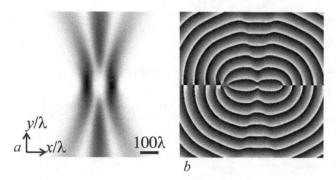

Fig. 3.24. Intensity (*a*) and phase (*b*) the sum of two identical asymmetric modes of Bessel and the first and second types (3.106). The calculation parameters are the same as in Fig. 3.23, but $c = 7$.

of the folding modes. Figure 3.23 shows the intensity (*a*) and phase (*b*) of the sum (3.106). The calculation parameters are the same as for Fig. 3.20.

With an increase in the asymmetry parameter ($c = 7$), both crescents, which, in their intensity section, form each of the beams described by the terms in (3.106), are transformed into astigmatic (elongated in the vertical coordinate) Gaussian beams and are shifted further from each other, forming a symmetrical intensity distribution, similar to the letter Ж (Russian alphbet) (Fig. 3.24 *a*). Figure 3.24 shows the intensity (*a*) and phase (*b*) of the sum of two asymmetric Bessel modes of different types (3.106) with $c = 7$ (other parameters, as in Fig. 3.20).

Note that the size of all images in Fig. 3.20–Fig. 3.23 is 200λ, and only the size of pictures in Fig. 3.24 is 400λ. The difference of two asymmetric Bessel modes of the first and second types

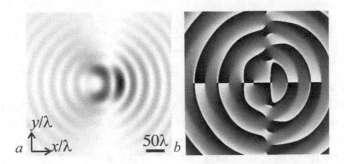

Fig. 3.25. Intensity (*a*) and phase (*b*) of the difference between two identical asymmetric Bessel modes of the first and second types (3.107). The calculation parameters are the same as in Fig. 3.23

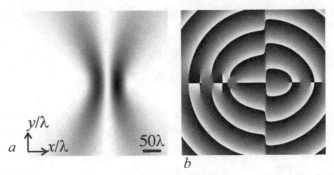

Fig. 3.26. Intensity (*a*) and phase (*b*) of the difference of two identical asymmetric Bessel modes of the first and second types (3.107). The calculation parameters are the same as in Fig. 3.25, but $c = 3$.

$$E_{1-2}(r,\varphi,z) = E_1(r,\varphi,z;c) - E_2(r,\varphi,z;c) \qquad (3.107)$$

for small values of the asymmetry parameter ($c = 1$), it forms an asymmetric intensity pattern (Fig. 3.25 *a*), but more complicated than the intensity picture for the sum of the same beams (Fig. 3.23).

From Fig. 3.25 *b* it can be seen that optical vortices with alternating topological charges of +1 and −1 are formed along the central vertical line.

If we increase the asymmetry parameter ($c = 3$), then the picture of the intensity of the difference between the two beams (3.107) will already be a more symmetrical form about the Cartesian axes, and will be similar to the Russian letter Ж (Fig. 3.26). Figure 3.26 shows the intensity (*a*) and phase (*b*) of the difference between two identical asymmetric modes of Bessel and the first and second types (3.107). The calculation parameters are the same as in Fig. 3.25, but $c = 3$. The size of the pictures in Figs. 3.25 and 3.26 is 200λ. Figure 3.26 shows that the intensity along the vertical axis y is zero (linear dislocation).

Figures 3.25 *b* and 3.26 *b* show that in the centre of the pictures of intensity for the difference of two identical asymmetric Bessel beams of the first and second type in the middle there is no increase in the unit topological charge of the optical vortex. In Figs 3.25 *b* and 3.26 *b* the topological charge of the optical vortex in the centre of the picture is equal to $n = 3$, as in the original beams, included in the difference (3.107).

Next we consider a complex displacement of the Bessel mode argument (3.91) of a more general form

$$\begin{cases} y = y' + ic/\alpha, \\ x = x' + b/\alpha. \end{cases} \tag{3.108}$$

In this case, a two-parameter (c, b) family of the asymmetric Bessel modes is obtained, in contrast to the one-parametre family of modes (3.90):

$$E_3(r,\varphi,c,b) = \left[\alpha r + (c-b)\exp(-i\varphi)\right]^n \times$$
$$\times \left[\alpha^2 r^2 - 2\alpha r(b\cos\varphi + ic\sin\varphi) + (b^2 - c^2)\right]^{-n/2} \times \tag{3.109}$$
$$\times J_n\left\{\sqrt{\alpha^2 r^2 - 2\alpha r(b\cos\varphi + ic\sin\varphi) + (b^2 - c^2)}\right\}\exp(in\varphi).$$

When $b = c$, expression (3.109) coincides with (3.90). The modes (3.109) are convenient because one can continuously change their shape by changing the parameter c, so that the mode does not shift $(b = 0)$. For the modes (3.90) and (3.97), with a change in the cross section of the intensity of the mode it automatically shifts along the x axis.

The asymmetric Bessel modes of the first type (3.90) can be expanded in a series in the unshifted Bessel modes (3.91) [87, 88]:

$$E_1(r,\varphi,c) = \left[\frac{\alpha r - 2c\exp(i\varphi)}{\alpha r}\right]^{-n/2} J_n\left\{\sqrt{\alpha r(\alpha r - 2c\exp(i\varphi))}\right\} \times$$
$$\times \exp(in\varphi) = \sum_{p=0}^{\infty}\frac{c^p \exp\left[i(n+p)\varphi\right]}{p!}J_{n+p}(\alpha r), \tag{3.110}$$

and the asymmetric Bessel modes of the second type (3.97) can also be expanded in a series of the Bessel unshifted modes (3.91):

$$
E_2\left(r,\varphi,c\right)=\left[\frac{\alpha r+2c\exp\left(-i\varphi\right)}{\alpha r}\right]^{n/2} J_n\left\{\sqrt{\alpha r\left(\alpha r+2c\exp\left(-i\varphi\right)\right)}\right\}\times
$$

$$
\times\exp\left(in\varphi\right)=\sum_{p=0}^{\infty}\frac{c^p\exp\left[i\left(n-p\right)\varphi\right]}{p!}J_{n-p}\left(\alpha r\right). \tag{3.111}
$$

The expressions (3.110) and (3.111) are used to calculate the OAM (3.99), to calculate the spatial spectrum of modes (3.90) and (3.97), and the scalar product of two modes with different parameters [87, 88]. For the two-parameter modes (3.109), it is not possible to obtain expansion in a series in the non-shifted Bessel modes, therefore, the OAM of such modes is difficult to calculate.

3.4.3. Orbital angular momentum of superpositions of asymmetric Bessel modes of the first and second type

Here we consider a superposition of modes of the first type with arbitrary complex coefficients:

$$
E_1\left(r,\varphi,z;c\right)=C_1E_1\left(r,\varphi,z;c\right)+C_2E_1\left(r,\varphi,z;-c\right). \tag{3.112}
$$

Let us determine the orbital angular momentum J_z (the projection of the OAM on the optical axis) and the total intensity I of the light beam in a plane transverse to the optical axis in the same way as in [88], i.e. using expressions for complex amplitudes in the form of a series of Bessel functions (3.110), (3.111). Then the OAM will look like this:

$$
J_z=2\pi D_1\lim_{R\to\infty}\sum_{p=0}^{\infty}\frac{\left(n+p\right)\left|c\right|^{2p}}{\left(p!\right)^2}\int_0^R J_{n+p}^2\left(\alpha r\right)rdr+
$$

$$
2\pi D_2\lim_{R\to\infty}\sum_{p=0}^{\infty}\left(-1\right)^p\frac{\left(n+p\right)\left|c\right|^{2p}}{\left(p!\right)^2}\int_0^R J_{n+p}^2\left(\alpha r\right)rdr. \tag{3.113}
$$

where $D_1=\left|C_1\right|^2+\left|C_2\right|^2$, $D_2=C_1^*C_2+C_1C_2^*$.

Similarly, the total energy is equal to:

$$I = 2\pi D_1 \lim_{R\to\infty} \sum_{p=0}^{\infty} \frac{|c|^{2p}}{(p!)^2} \int_0^R J_{n+p}^2(\alpha r)\, r\, dr +$$

$$+2\pi D_2 \lim_{R\to\infty} \sum_{p=0}^{\infty} (-1)^p \frac{|c|^{2p}}{(p!)^2} \int_0^R J_{n+p}^2(\alpha r)\, r\, dr.$$

(3.114)

Since for any integer m the integrals in the sums for large values of R are equal

$$\int_0^R J_m^2(\alpha r)\, r\, dr = \frac{R}{\pi\alpha},$$

one can get the expression for the normalized OAM of the light field (3.112):

$$\frac{J_z}{I} = n + |c| \frac{D_1 I_1(2|c|) + D_2 J_1(2|c|)}{D_1 I_0(2|c|) + D_2 J_0(2|c|)}.$$

(3.115)

From (3.115) it follows that for the superposition of two modes of the first type without a phase difference between them (i.e, $C_1 = C_2 = 1$, (3.113)) the normalized OAM is equal to:

$$\frac{J_z}{I} = n + |c| \frac{I_1(2|c|) + J_1(2|c|)}{I_0(2|c|) + J_0(2|c|)}.$$

(3.116)

If, in the superposition of two modes of the first type, the phase diffeence is π (that is, $C_1 = 1$, $C_2 = -1$, (3.104)) the normalized OAM is equal to:

$$\frac{J_z}{I} = n + |c| \frac{I_1(2|c|) - J_1(2|c|)}{I_0(2|c|) - J_0(2|c|)}.$$

(3.117)

We now consider the superposition of modes of different types with arbitrary complex coefficients:

$$E_{12}(r,\varphi,z;c) = C_1 E_1(r,\varphi,z;c) + C_2 E_2(r,\varphi,z;c).$$

(3.118)

In this case, the OAM and the total intensity are:

$$J_z = 2\pi |C_1|^2 \lim_{R\to\infty} \sum_{p=0}^{\infty} \frac{(n+p)|c|^{2p}}{(p!)^2} \int_0^R J_{n+p}^2(\alpha r) r \, dr + 2\pi |C_2|^2 \times$$

$$\times \lim_{R\to\infty} \sum_{p=0}^{\infty} \frac{(n-p)|c|^{2p}}{(p!)^2} \int_0^R J_{n-p}^2(\alpha r) r \, dr + \tag{3.119}$$

$$+ 2\pi n \left(C_1^* C_2 + C_1 C_2^* \right) \lim_{R\to\infty} \int_0^R J_n^2(\alpha r) r \, dr,$$

$$I = 2\pi |C_1|^2 \lim_{R\to\infty} \sum_{p=0}^{\infty} \frac{|c|^{2p}}{(p!)^2} \int_0^R J_{n+p}^2(\alpha r) r \, dr +$$

$$+ 2\pi |C_2|^2 \lim_{R\to\infty} \sum_{p=0}^{\infty} \frac{|c|^{2p}}{(p!)^2} \int_0^R J_{n-p}^2(\alpha r) r \, dr + \tag{3.120}$$

$$+ 2\pi \left(C_1^* C_2 + C_1 C_2^* \right) \lim_{R\to\infty} \int_0^R J_n^2(\alpha r) r \, dr.$$

From (3.119), (3.120) it is possible to obtain the expression for the normalized OAM:

$$\frac{J_z}{I} = n + \frac{\left(|C_1|^2 - |C_2|^2 \right) |c| I_1(2|c|)}{\left(|C_1|^2 + |C_2|^2 \right) I_0(2|c|) + \left(C_1^* C_2 + C_1 C_2^* \right)}. \tag{3.121}$$

In particular, (3.121) shows that in case of the sum (3.106) and the difference (3.107) the OAM is independent of α and c, and is equal to n.

Expression (3.121) can be considered the main result of this work, since (3.121) shows that by the addition and subtraction of two aB-modes of the types I and II with the same n, α and c, it is possible, by changing the parameters α and c, to obtain modes with a different distribution of transverse intensity, but with the same OAM, equal to n. Therefore, it turns out that the modes in Figs. 3.21–3.26 all have the same OAM, equal to $n = 3$.

Thus, it is theoretically and numerically shown in this section [158] that the previously considered asymmetric Bessel [87, 88] and Bessel–Gauss [124, 155] modes are obtained by complex displacement of the ordinary Bessel mode. It was also shown that a complex displacement of a Bessel mode of another type can form an asymmetric Bessel mode of the second type (in relation to modes of the first type [88]). The asymmetric Bessel modes of the first and second types have a different orbital angular momentum. By combining together the asymmetric Bessel modes of both types in the form of a sum or difference, we can also obtain Bessel modes, but with different intensity distributions in the cross section. In particular, it is possible to obtain symmetric Bessel modes similar to the letters O, Ж, X. It is also interesting that the difference between two identical mirror asymmetric Bessel modes with topological charges n leads to the formation of an optical vortex with a topological charge $n + 1$ in the centre. A difference or the sum of two asymmetric Bessel modes of the first and second type with the same parameters has an OAM equal to the topological charge n of the original (generating) Bessel mode and independent of other parameters: the scale α of the mode and the asymmetry parameter c.

3.5. Theorems on the orbital angular momentum

Laser beams with an orbital angular momentum (OAM) have found application in micro-object manipulation, quantum telecommunications, microscopy, interferometry, and metrology. A recent review of papers on OAM is given in [101]. In [9], it was first shown that the Laguerre–Gauss modes have OAMs. Among the beams that possess OAM, the Bessel beams are also known that can propagate without diffraction. In some papers, analytical expressions were obtained for the density of an OAM for Bessel beams [130, 131, 150]. Recently, nonparaxial asymmetric Bessel modes [88] were considered, for which an analytical expression was also obtained for the OAM of the entire beam. The superposition of unshifted vortex laser beams, in particular Bessel beams, was considered previously [150, 132–134, 141], and the work [159] examined the OAM of the superposition of vortex laser beams with periodic displacement. In [160], the formation of paraxial light beams of various shapes with a given value of the orbital angular momentum is considered.

In this section we discuss the superposition of arbitrary identical radially symmetric optical vortices (including nonparaxial), each of which is shifted by an arbitrary vector in a plane transverse to the optical axis. Two theorems are proved. The first theorem states that if the superposition weighting coefficients are real (i.e., the phase difference between the beams in the superposition is 0 or π), then the normalized OAM (more precisely, the projection of the OAM on the optical axis) of the entire superposition is equal to the normalized

OAM of each beam entering it. The second theorem states that in order to preserve the normalized OAM, the superposition coefficients may be not real, but then the centres of all the vortices must be on one straight line passing through the origin (the point relative to which the OAM is calculated). The numerical calculation of the OAM for the superpositions of three and five optical vortices of different shapes confirms the assertions of both theorems.

3.5.1. Orbital angular momentum of superpositions of identical optical vortices with radial symmetry

It is well known that any solution of the nonparaxial Helmholtz equation can be represented as a superposition of plane waves:

$$E(r,\phi,z) = \iint_{\mathbb{R}^2} A(\rho,\theta) \exp\left[ikr\rho\cos(\theta-\phi) + ikz\sqrt{1-\rho^2} \right] \rho\,d\rho\,d\theta, \quad (3.122)$$

where $k = 2\pi/\lambda$ is the wave number of light with wavelength λ, E is the complex amplitude field in the cylindrical coordinates (ρ, φ, z), A is the angular spectrum in the polar coordinates (ρ, θ). In the paraxial approximation instead of the root $(1 - \rho^2)^{1/2}$ there will be will be $(1 - \rho^2/2)$ in the exponent.

If the original beam is shifted by a vector with Cartesian coordinates (x_0, y_0), then its angular spectrum of plane waves has the form:

$$A'(\rho,\theta) = A(\rho,\theta) \exp\left[-ik\rho(x_0\cos\theta + y_0\sin\theta) \right], \quad (3.123)$$

where $A(\rho, \theta)$ is the angular spectrum of plane waves of the original (unshifted) beam. The OAM and the beam power can be calculated through both the complex amplitude and the angular spectrum of plane waves. One can show the projection of the OAM on the optical axis J_z and the power of the laser beam I is calculated using the relations:

$$J_z = -i\iint_{\mathbb{R}^2} E^* \frac{\partial E}{\partial\phi} r\,dr\,d\phi = -i\lambda^2 \iint_{\mathbb{R}^2} A^* \frac{\partial A}{\partial\theta} \rho\,d\rho\,d\theta, \quad (3.124)$$

$$I = \iint_{\mathbb{R}^2} E^* E r\,dr\,d\phi = \lambda^2 \iint_{\mathbb{R}^2} A^* A \rho\,d\rho\,d\theta. \quad (3.125)$$

Normalized OAM (OAM per photon) is obtained as the ratio of J_z/I.
Let the angular spectrum of plane waves of a vortex laser beam be:

$$A(\rho,\theta) = D(\rho)\exp(in\theta). \tag{3.126}$$

where $D(\rho)$ is an arbitrary, generally complex-valued, function, n is
an integer topological charge. Then the complex amplitude of such
a vortex beam will look like:

$$E(r,\phi,z) = \iint_{\mathbb{R}^2} D(\rho)\exp(in\theta)\exp\left[ikr\rho\cos(\theta-\phi) + ikz\sqrt{1-\rho^2}\right] \times$$

$$\times \rho\, d\rho\, d\theta = 2\pi i^n \exp(in\phi) \times \tag{3.127}$$

$$\times \int_0^\infty D(\rho)\exp\left(ikz\sqrt{1-\rho^2}\right) J_n(kr\rho)\rho\, d\rho = B(r,z)\exp(in\phi).$$

And in the initial plane, the complex amplitude will look like
(3.126):

$$E(r,\phi,z=0) = B(r,z=0)\exp(in\phi) = B(r)\exp(in\phi). \tag{3.128}$$

Consider a superposition of M identical optical vortices (3.128)
with weighting coefficients C_m, with each beam shifted in the
Cartesian plane by a vector with coordinates (x_m, y_m), $m = 0, ..., M-1$.
The angular spectrum of plane waves of the entire superposition is
equal to

$$A(\rho,\theta) = D(\rho)\exp(in\theta)\sum_{m=0}^{M-1} C_m \exp\left[-ik\rho(x_m\cos\theta + y_m\sin\theta)\right]. \tag{3.129}$$

Substituting (3.128) into (3.126), we calculate the power of such a beam:

$$I = \lambda^2 \sum_{p=0}^{M-1}\sum_{q=0}^{M-1} C_p^* C_q \int_0^\infty |D(\rho)|^2 \rho\, d\rho \times$$

$$\int_0^{2\pi} \exp\left[ik\rho(a_{pq}\cos\theta + b_{pq}\sin\theta)\right]d\theta, \tag{3.130}$$

where $a_{pq} = x_p - x_q$, $b_{pq} = y_p - y_q$.
Similarly, let's calculate the OAM (3.124):

$$J_z = nI + \lambda^2 \sum_{p=0}^{M-1}\sum_{q=0}^{M-1} S_{pq}, \tag{3.131}$$

where

$$
S_{pq} = kC_p^* C_q \int\limits_0^\infty |D(\rho)|^2 \, \rho^2 d\rho \int\limits_0^{2\pi} \left(x_q \sin\theta - y_q \cos\theta\right) \times
$$

$$
\times \exp\left[ik\rho\left(a_{pq} \cos\theta + b_{pq} \sin\theta\right)\right] d\theta. \tag{3.132}
$$

Dividing the OAM by the beam power, we obtain the expression for the normalized OAM:

$$
\frac{J_z}{I} = n + \frac{\lambda^2}{I} \sum_{p=0}^{M-1} \sum_{q=0}^{M-1} S_{pq}, \tag{3.133}
$$

First, in the double sum in (3.133), we calculate the diagonal terms (i.e., with $p = q$)

$$
S_{pp} = k \left|C_p\right|^2 \int\limits_0^\infty |D(\rho)|^2 \, \rho^2 d\rho \int\limits_0^{2\pi} \left(x_p \sin\theta - y_p \cos\theta\right) d\theta = 0. \tag{3.134}
$$

The off-diagonal terms are included in the double sum in (3.133) in pairs, we calculate the sum of two symmetric terms:

$$
S_{pq} + S_{qp} = kC_p^* C_q \int\limits_0^\infty |D(\rho)|^2 \, \rho^2 d\rho \int\limits_0^{2\pi} \left(x_q \sin\theta - y_q \cos\theta\right) \times
$$

$$
\times \exp\left[ik\rho\left(a_{pq} \cos\theta + b_{pq} \sin\theta\right)\right] d\theta + kC_p C_q^* \int\limits_0^\infty |D(\rho)|^2 \, \rho^2 d\rho \tag{3.135}
$$

$$
\times \int\limits_0^{2\pi} \left(x_p \sin\theta - y_p \cos\theta\right) \times \exp\left[ik\rho\left(-a_{pq} \cos\theta - b_{pq} \sin\theta\right)\right] d\theta.
$$

In the second integral we make the change $\theta \rightarrow \theta + \pi$, but the integration limits remain the same as the integration of a periodic function throughout the period. Then the exponent arguments in both integrals are the same and the expression (3.135) takes the form:

$$
S_{pq} + S_{qp} = k \int\limits_0^\infty |D(\rho)|^2 \, \rho^2 d\rho \int\limits_0^{2\pi} \left[\left(C_p C_q^* y_p - C_p^* C_q y_q\right)\cos\theta -
$$

$$
-\left(C_p C_q^* x_p - C_p^* C_q x_q\right)\sin\theta \right] \exp\left[ik\rho\left(a_{pq} \cos\theta + b_{pq} \sin\theta\right)\right] d\theta. \tag{3.136}
$$

If all the weight coefficients of the superposition (3.129) are real (i.e., the phase difference between the beams in the superposition is 0 or π), then the integrand can be represented as the total differential with respect to the variable θ of the exponent, i.e.

$$S_{pq} + S_{qp} = -iC_p C_q \int_0^{\infty} |D(\rho)|^2 \rho d\rho \times$$

$$\times \exp\left[ik\rho\left(a_{pq}\cos\theta + b_{pq}\sin\theta\right)\right]\Big|_{\theta=0}^{\theta=2\pi} = 0,$$

(3.137)

therefore, the second term in (3.133) is zero. Thus, we have proved the following theorem.

Theorem 1. Let there be a light beam that is a superposition of identical optical vortices shifted from the optical axis with an arbitrary radial shape and topological charge n, all the weight coefficients of the superposition being real, and the displacement of each beam can be arbitrary. Then the normalized OAM of the entire superposition is equal to the OAM of one beam $J_z/I = n$.

As a special case, it follows from the theorem that the displacement of a single beam in the transverse plane does not lead to a change in its normalized OAM [159].

If the coefficients in the superposition (3.129) are complex, then the following integral can be used to calculate the normalized OAM:

$$\int_0^{2\pi} \exp(ia\cos\theta + ib\sin\theta)\begin{Bmatrix} \cos\theta \\ \sin\theta \end{Bmatrix} d\theta =$$

$$= \frac{i2\pi}{\sqrt{a^2+b^2}}\begin{Bmatrix} a \\ b \end{Bmatrix} J_1\left(\sqrt{a^2+b^2}\right).$$

(3.138)

Then, from (3.136) instead of (3.137) we get:

$$S_{pq} + S_{qp} = \frac{2\pi ik}{R_{pq}}\Big[\left(C_p C_q^* y_p - C_p^* C_q y_q\right)\left(x_p - x_q\right) -$$

$$-\left(üüjiüjiü_p - _p^* _q _q\right)\left(_p - _q\right)\Big]\int_0^{\infty} |(\rho)|^2 {}_1\left({}_{pq}\rho\right)\rho^2 \rho,$$

(3.139)

where $R_{pq} = [(x_p - x_q)^2 + (y_p - y_q)^2]^{1/2}$ is the distance between the centres of the p-th and q-th vortices. In order for the sum (3.139) to be equal to zero, it is necessary that the following condition be satisfied:

$$C_p C_q^* \left[y_p \left(x_p - x_q \right) - x_p \left(y_p - y_q \right) \right] =$$
$$= C_p^* C_q \left[y_q \left(x_p - x_q \right) - x_q \left(y_p - y_q \right) \right], \tag{3.140}$$

which, after the abbreviations in square brackets, is given to the form:

$$C_p C_q^* \left(x_p y_q - x_q y_p \right) = C_p^* C_q \left(x_p y_q - x_q y_p \right). \tag{3.141}$$

If the coefficients C_p and C_q are not real, then condition (3.141) is satisfied only if the expressions in parentheses are zero. Thus, we have proved the following theorem.

Theorem 2. Let there be a light beam that is a superposition of M identical optical vortices shifted from the optical axis with an arbitrary radial shape and topological charge n, and the centres of all optical vortices (x_p, y_p) $(p = 0,..., M - 1)$ in superposition, they are located on one straight line passing through the origin of coordinates (ie, $x_p y_q = x_q y_p$ for any p and q). Then the normalized OAM of the entire superposition is equal to the OAM of one beam $J_z/I = n$.

3.5.2. Numerical simulation

Verification of the statements of both theorems was carried out by numerical simulation. Four different types of optical vortices with a topological charge n were considered: Bessel diffraction modes [125], Hankel–Bessel nonparaxial beams [146], Bessel–Gauss paraxial beams [119], and Laguerre–Gauss paraxial modes [9]. The complex amplitude of the Bessel beam in the initial plane is [125]:

$$E_1(r, \varphi, z = 0) = J_n(\alpha r) \exp(in\varphi), \tag{3.142}$$

where α is a scaling factor. The complex amplitude of the Hankel-Bessel beam in the initial plane is [146]:

$$E_2(r, \varphi, z = 0) = H_{n/2}^{(1)} \left(\frac{kr}{2} \right) J_{n/2} \left(\frac{kr}{2} \right) \exp(in\varphi), \tag{3.143}$$

where k is the wave number of light. The complex amplitude of the Bessel–Gauss beam in the initial plane is [119]:

$$E_3(r,\varphi,z=0) = J_n(\alpha r)\exp\left(-\frac{r^2}{w^2}+in\varphi\right),\tag{3.144}$$

where α is the scaling factor, w is the radius of the Gaussian beam waist. The complex amplitude of the Laguerre–Gauss beam in the initial plane is equal to [9]:

$$E_4(r,\varphi,z=0) = \left(\frac{r\sqrt{2}}{w}\right)^n L_m^n\left(\frac{2r^2}{w^2}\right)\exp\left(-\frac{r^2}{w^2}+in\varphi\right),\tag{3.145}$$

where m is the order of the mode, w is the radius of the Gaussian beam waist, $L_m^n(x)$ is the Laguerre polynomial.

Figure 3.27 shows the intensity and phase of superpositions of three beams, the centres of which are located at the vertices of a regular triangle. All beams in superpositions have a topological charge $n = 7$. Figure 3.27 was calculated using the formulas (3.142)–(3.145) with the number of samples 1000×1000. The parameters of the beams were chosen so that the light triangle in all diffraction patterns had approximately the same dimensions. The normalized OAM of all four beams was calculated using the formulas (3.124), (3.125) and turned out to be equal $J_z/I \cong 7$, despite the difference in the diffraction patterns. Similarly, in Fig. 3.27 the intensities and phases of superpositions are shown from five of the same beams, but located at the vertices of a regular pentagon (larger). The normalized OAM of all beams in Fig. 3.27 was also calculated using the formulas (3.124), (3.125) and also turned out to be equal $J_z/I \cong 7$.

Figure 3.29 a, b shows the intensity and phase of the superposition of five Laguerre–Gauss modes of order $(n, m) = (7, 2)$ located on the straight line $y = 0.4x$ at the points with coordinates $x_m = [-12, -6, 0, 6, 12]\lambda$, $y_m = 0.4x_m$, $m = 0, ..., 4$, with parameters $\lambda = 532$ nm, $w = \lambda$, weight coefficient vector $C_m = \exp(2\pi im/5)$, the size of the computational domain $2R = 40\lambda$ (i.e. $-R \leq x, y \leq R$).

Figure 3.29 c, d shows the intensity and phase of a superposition of five 7th-order Bessel–Gauss beams, which are located at the same points as in Fig. 3.29 a, b (i.e., on the line $y = 0.4x$ at the points with coordinates $x_m = [-12, -6, 0, 6, 12]\lambda$, $y_m = 0.4x_m$, $m = 0, ..., 4$), with parameters $\lambda = 532$ nm, $w = 2\lambda$, $\alpha = 2/\lambda$, the vector of weighting coefficients $C_m = \exp(2\pi im/5)$, the size of the computational domain $2R = 40\lambda$ (i.e. $-R \leq x, y \leq R$).

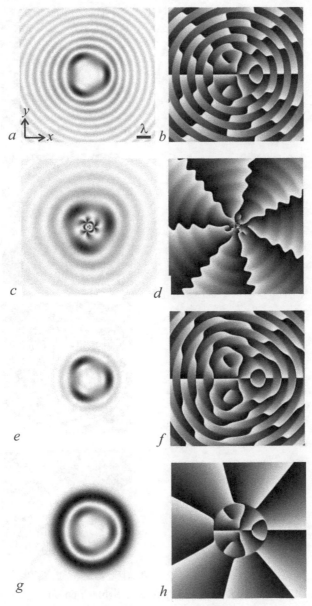

Fig. 3.27. Intensity (negative) (*a, b, d, g*) and phase (black colour – π, white colour – $-\pi$) (*b, d, e, h*) of superpositions of three vortex beams (wavelength $\lambda = 532$ nm), arranged in an equilateral triangle vertices at coordinates $x_p = R_0 \cos \varphi_p$, $y_p = R_0 \sin \varphi_p$ ($R_0 = 0.3\lambda$, $\varphi_p = 2\pi p/3$, $p = 0, 1, 2$), and having a topological charge equal to 7: Bessel modes (*a, b*), Hankel–Bessel beams (*c, d*), Bessel–Gauss beams (*e, f*) and Laguerre-Gauss beams (*g, h*). Weight vector in each of superposition is $\mathbf{C} = (1, 1, 1)$, the size of the computational domain $2R = 10\lambda$. Scaling factor of the Bessel and Bessel–Gauss beams (*a, b, e, f*) $\alpha = 6/\lambda$. Gaussian beam waist radius equal to $w = 2\lambda$ (*e, f*) and $w = \lambda$(*g, h*). Laguerre-Gauss mode index (*g, h*): $(n, m) = (7, 1)$.

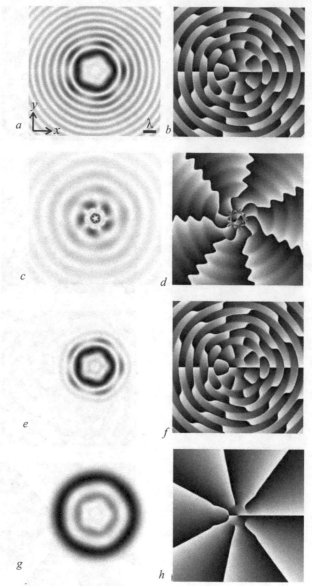

Fig. 3.28. Intensity (negative) (*a, c, e, g*) and phase (black colour – π, white colour – $-\pi$) (*b, d, f, h*) of superpositions of five vortex beams (wavelength $\lambda = 532$ nm), located at the vertices of a regular pentagon at coordinates $x_p = R_0 \cos \varphi_p$, $y_p = R_0 \sin \varphi_p$ ($R_0 = 0.5\lambda$, $\varphi_p = 2\pi p/5$, $p = 0, ..., 4$), and having a topological charge equal to 7: Bessel modes (*a, b*), Hankel–Bessel beams (*c, d*), Bessel–Gauss beams (*e, f*) and Laguerre–Gauss beams (*g, h*). Weight vector in each of superposition is $\mathbf{C} = (1, 1, 1, 1, 1)$, the size of the computational domain $2R = 10\lambda$. Scaling factor of the Bessel and Bessel–Gauss beams (*a, b, e, f*) $\alpha = 6/\lambda$. Gaussian beam waist radius equal to $w = 2\lambda$ (*e, f*) and $w = \lambda$ (*g, h*). Laguerre–Gauss mode index (*g, h*): $(n, m) = (7, 1)$.

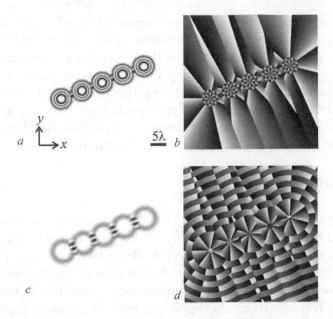

Fig. 3.29. Intensity (negative) (*a, c*) and phase (black colour – π, white colour – $-\pi$ (*b, d*) of superposition of five Laguerre–Gauss modes (*a, b*) and Bessel–Gauss beams (*c, d*). Laguerre–Gaussian mode parameters: the order $(n, m) = (7, 2)$, waist radius $w = \lambda$. Parameters of the Bessel–Gauss beams: the order $n = 7$, the waist radius $w = 2\lambda$, the scaling factor $\alpha = 2/\lambda$. All the beams have a wavelength $\lambda = 532$ nm, the centres are located on the straight line $y = 0.4x$ at the points with coordinates $x_m = [-12, -6, 0, 6, 12]\lambda$, $y_m = 0.4x_m$, $m = 0,..., 4$, the vector of weighting coefficients in superposition $C_m = \exp(2\pi i m/5)$, the size of the computational domain $2R = 40\lambda$.

The normalized OAM of both beams in Fig. 3.29 was also calculated using the formulas (3.124), (3.125) and also turned out to be equal $J_z/I \cong 7$.

Thus, in this section attention is given to the OAM of the superposition of identical radially symmetric optical vortices, each of which is shifted in a plane transverse to the optical axis [161]. Two theorems on the preservation of the normalized OAM of such a superposition are proved. If the superposition weighting coefficients are real (i.e., the phase difference between the beams is 0 or π), then, irrespective of the displacement of the beams, the normalized OAM of the whole superposition is equal to the normalized OAM of each beam entering into it, i.e. its topological charge. If the centres of the beams are located on one straight line passing through the origin of coordinates (the point relative to which the OAM is calculated), then even with the complex weight coefficients, the normalized OAM of the whole superposition is also equal to the normalized

OAM of each beam entering it. A numerical simulation was carried out, which confirmed the conservation of the normalized OAM for superpositions of three and five optical vortices of different shapes. The proved theorems allow using any number of identical radially symmetric vortex beams, by selecting the weighting factors and the magnitude of displacement of each beam from the optical axis, to form new beams with different intensity distribution (including not radially symmetric), but with the same normalized OAM.

3.6. Lommel modes

Diffraction-free beams occupy a special place among the known laser beams. The structure of the distribution of their complex amplitude in the cross section is such that, despite the diffraction, it is preserved with the propagation of an arbitrary distance along the optical axis. It is known that in the three-dimensional space the Bessel modes [1] are diffraction-free, whereas in the two-dimensional space – Airy beams [18]. Also, diffraction-free beams were generalized to the case of a higher dimension of space [162]. It is also known that in three-dimensional space a light beam is diffraction-free, for which the angular spectrum of plane waves is an infinitely thin circle. So, in [118] the diffraction-free beams, described as a linear combination of Bessel modes, are considered. The complex amplitude of such beams is described by the Mathieu function. Work [87] deals with asymmetric Bessel modes whose intensity distribution in the cross section has the form of a crescent, and in [88] a generalization of this set is considered by introducing an additional parameter that allows controlling the asymmetry of the transverse intensity distribution. Diffraction-free beams are stable during propagation in a turbulent atmosphere [163] and femtosecond Bessel pulses retain their shape during propagation [164].

In this section we consider a linear combination of Bessel modes with such coefficients that the complex amplitude of the beam is described by the Lommel function of two variables, one of which is complex. The Lommel functions of two variables are found in optics not for the first time. Thus, in [59] (section 8.8) the three-dimensional distribution of light near the focus was considered for a spherical monochromatic wave emerging from a circular aperture and converging at an axial focal point. To obtain this distribution, the Fresnel integral is used in [59], expressed in terms of the Lommel functions of two variables [1 65]. In recent work [166] using these functions, the focusing was described using a Laguerre–Gauss vortex laser lens with a zero radial index limited by a circular aperture. Unlike the traditional Bessel modes [1], the intensity distribution of the Lommel mode (L-mode) does not have a radially symmetric shape as a set of light rings, and, unlike asymmetric modes from [87, 88], it has a symmetry with respect to more than one or both Cartesian axes. Below the orbital angular momentum of the L-beams is precisely calculated. It exceeds the momentum of the Bessel mode, which is included in the linear combination with the lowest topological charge. The L-modes, like all diffraction-free beams, have infinite energy, and therefore can only be realized in practice only approximately. We call these beams Lommel beams by analogy with the Bessel beams. But since the Bessel beams are modes, the Lommel beams are also sometimes called the Lommel modes.

3.6.1. Complex amplitude of Lommel beams

Here we consider a light beam whose angular spectrum of plane waves has the following form:

$$A(\rho,\theta) = \frac{(-i)^n}{\lambda\alpha}\delta\left(\rho-\frac{\alpha}{k}\right)\sum_{p=0}^{\infty}c^{2p}\exp\left[i(n+2p)\theta\right] =$$
$$= \frac{(-i)^n\exp(in\theta)}{\lambda\alpha\left[1-c^2\exp(2i\theta)\right]}\delta\left(\rho-\frac{\alpha}{k}\right), \tag{3.146}$$

wherein (ρ, θ) are the polar coordinates in the spectral plane, $\delta(x)$ is the Dirac delta function, $k = 2\pi/\lambda$ is the wave number of light with wavelength λ, and the parameters α, c and n, as will be shown below, characterize respectively the scale of the beam, the asymmetry of its shape and the orbital angular momentum. We note that, in contrast to the asymmetric Bessel modes from [87, 88], the asymmetry parameter c for L-beams cannot be arbitrary, but must be less than one in absolute value, otherwise the series in (3.147) will be divergent. From (3.146) it can be seen that the modulus of the amplitude of the spectrum varies along a ring with a radius of $\rho = \alpha/k$: for real values of c, the maximum value takes place at $\theta = 0, \pi$, and the minimum – at $\theta = \pm\pi/2$. The complex amplitude of the L-beam is located as the Fourier transform of the angular spectrum (3.146) and is equal to

$$E_n(r,\varphi,z) = \exp\left(iz\sqrt{k^2 - \alpha^2}\right) =$$

$$= \sum_{p=0}^{\infty} (-1)^p c^{2p} \exp[i(n+2p)\varphi] J_{n+2p}(\alpha r) = \qquad (3.147)$$

$$= c^{-n} \exp\left(iz\sqrt{k^2 - \alpha^2}\right) U_n[car\exp(i\varphi), \alpha r].$$

where $U_n(w, z)$ is the Lommel function of two variables [165]:

$$U_n(w,z) = \sum_{p=0}^{\infty} (-1)^p \left(\frac{w}{z}\right)^{n+2p} J_{n+2p}(z). \qquad (3.148)$$

In (3.146) parameter α is included in the argument of the Bessel functions, so it characterizes the scale (the width of the light ring) of the Lommel beam.

Using (3.146) it is easy to show that

$$\left|E_n(r,\varphi,z)\right| = \left|E_n(r,-\varphi,z)\right| = \left|E_n(r,\pi-\varphi,z)\right| =$$

$$= \left|\sum_{p=0}^{\infty} (-1)^p c^{2p} \exp[i(n+2p)\varphi] J_{n+2p}(\alpha r)\right|. \qquad (3.149)$$

It follows from (3.149) that, in contrast to the asymmetric Bessel beams from [87, 88], the intensity distribution in the cross section of the L-beams (3.147) is symmetrical not only with respect to the horizontal plane Oxz, but also with respect to the vertical plane Oyz.

At $c = 0$ Eq. (3.147) retains only one term and the L-beam is converted into a traditional Bessel beam $E_n(r,\varphi,z) = \exp\left(iz\sqrt{k^2 - \alpha^2} + in\varphi\right) J_n(\alpha r)$.

Figure 3.30 shows the intensity and phase distributions (at $z = 0$) in the transverse plane for L-beams with the following parameters: wavelength $\lambda = 532$ nm, topological charge $n = 4$, scaling factor $\alpha = k/3$, asymmetry parameter $c = 0.5i$ (*a, b*) and $c = 0.9i$ (*c, d*). Figure 3.30 shows the intensity in the region $-20\lambda \le x, y \le 20\lambda$. This figure was obtained by calculating using to the formula (3.147).

As can be seen from Fig. 3.30, for small values of the parameter c the diffraction pattern is similar to the picture of the Bessel mode, but extended along one Cartesian coordinate (*x*-axis). With a two-dimensional realization of such beams (that is, propagating in the Oxz plane), their shape will be similar to the shape of accelerating elliptic modes, considered in [149].

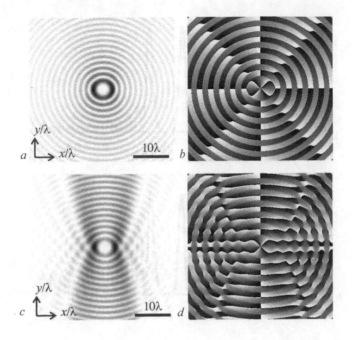

Fig. 3.30. The intensity distribution (negative) (*a, c*) and phase (*b, d*) in the transverse plane for the Lommel beams: c = 0.5*i* (*a, b*) and c = 0.9*i* (*b, c*).

Further, with an increase in the parameter *c*, the asymmetry of the L-beam increases, and in the cross section it has the form of two crescents with zero intensity in the centre. In optical micromanipulation, such a distribution is convenient for holding a micro-object in place along one coordinate [167].

Without loss of generality, parameter *c* can be considered real positive. Otherwise, the diffraction pattern is rotated by an angle corresponding to the argument of the parameter *c*. So, for Fig. 3.30 parameter *c* is purely imaginary, therefore the crescents are spaced along the *x* axis.

As the parameter *c* increases, not only does the asymmetry of the transverse intensity distribution increase, but also the intensity contrast along the Cartesian coordinate axes *x* and *y*. This is clearly seen from Fig. 3.31 calculated also by the formula (3.147) for the same parameters as Fig. 3.30.

From Fig. 3.31 it can be seen that if at c = 0.5*i* the maximum intensity in the horizontal plane exceeds the maximum intensity in the vertical plane by about one and a half times, then at c = 0.9*i* this

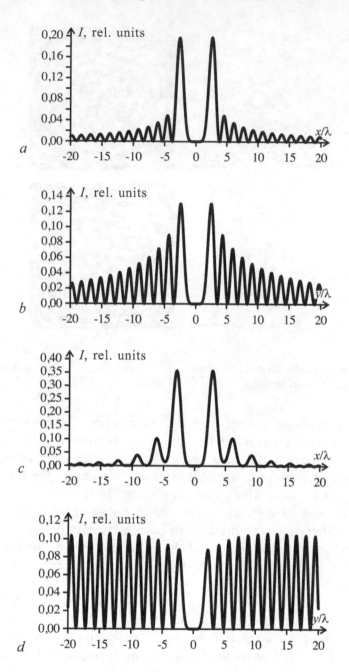

Fig. 3.31. The intensity cross sections at $y = 0$ (a, c) and $x = 0$ (b, d) for Lommel beams: $c = 0.5i$ (a, b) and $c = 0.9i$ (c, d). The horizontal axis shows the wavelengths, and the vertical intensity in relative units.

ratio is already 3.5. Moreover, in the vertical plane, the side lobes exceed the main central maxima.

3.6.2. Orbital angular momentum of Lommel beams

The orbital angular momentum (OAM) J_z (the projection of the OAM on the optical axis) and the total intensity I of the light beam in a plane transverse to the optical axis are determined by the following formulas [110]:

$$J_z = \mathrm{Im}\left\{ \iint_{\mathbb{R}^2} E^* \frac{\partial E}{\partial \varphi} r\, dr\, d\varphi \right\} = \mathrm{Im}\left\{ \lim_{R \to \infty} \int_0^R \int_0^{2\pi} E^* \frac{\partial E}{\partial \varphi} r\, dr\, d\varphi \right\}, \quad (3.150)$$

$$I = \iint_{\mathbb{R}^2} E^* E\, r\, dr\, d\varphi = \lim_{x \to \infty} \int_0^R \int_0^{2\pi} E^* E\, r\, dr\, d\varphi. \quad (3.151)$$

Substituting in (3.150), (3.151) the complex amplitude (3.146), we obtain the OAM and the total intensity of the Lommel beam:

$$J_z = 2\pi \lim_{x \to \infty} \sum_{p=0}^{\infty} (n + 2p)\left(cc^*\right)^{2p} \int_0^R J_{n+2p}^2(\alpha r)\, r\, dr. \quad (3.152)$$

$$I = 2\pi \lim_{x \to \infty} \sum_{p=0}^{\infty} \left(cc^*\right)^{2p} \int_0^R J_{n+2p}^2(\alpha r)\, r\, dr. \quad (3.153)$$

The integrals in these expressions are described in [112] (expression 5.54.2):

$$\int J_p^2(\alpha r)\, r\, dr = \frac{r^2}{2}\left[J_p^2(\alpha r) - J_{p-1}(\alpha r) J_{p+1}(\alpha r) \right]. \quad (3.154)$$

Using the asymptotic Bessel function for large values of the argument (expression 9.2.1 in [53]), we denote that all the integrals in (3.152) and (3.153) do not depend on the order of the Bessel functions and are equal to $R/(\pi\alpha)$. Then, using the numerical series 0.246.1 and 0.246.2 from [112] and the divisions (3.152) into (3.153), we obtain the expression for the OAM, normalized to the intensity:

$$\frac{J_z}{I} = \frac{\sum_{p=0}^{\infty}\left(cc^*\right)^{2p}\left(n+2p\right)}{\sum_{p=0}^{\infty}\left(cc^*\right)^{2p}} = n + \frac{2|c|^4}{1-|c|^4}. \tag{3.155}$$

From (3.155), it follows that with an increase in the asymmetry parameter $c < 1$ the OAM increases. In Fig. 3.30 *a, b* and for $c = 0.5i$, the normalized OAM is equal to $J_z/I \approx 4.1$, and in Fig. 3.30 *c, d* for $c = 0.9i$ it is $J_z/I \approx 7.8$.

3.6.3. Orthogonality of complex amplitudes of Lommel beams

Just as the orbital angular momentum was calculated in the previous section, we can calculate the scalar product of two Lommel beams – with topological charges n and m, scaling factors α and β, and c and d asymmetry parameters:

$$\left(E_{n\alpha c}, E_{m\beta d}\right) = \iint_{\mathbb{R}^2} E_{n\alpha c}E_{m\beta d}^* r\,dr\,d\varphi =$$

$$= \begin{cases} 2\pi(-1)^{(n-m)/2}\dfrac{\delta(\alpha-\beta)}{\alpha}\dfrac{\left(d^*\right)^{n-m}}{1-\left(cd^*\right)^2}, & \text{if } (m+n)\text{ is even and } n \geq m, \\[3mm] 2\pi(-1)^{(n-m)/2}\dfrac{\delta(\alpha-\beta)}{\alpha}\dfrac{c^{m-n}}{1-\left(cd^*\right)^2}, & \text{if } (m+n)\text{ is even and } n \leq m, \\[3mm] 0, & \text{if } (m+n)\text{ is odd.} \end{cases} \tag{3.156}$$

It can be seen from (3.156) that the complex amplitudes of L-beams, like the traditional and asymmetric Bessel beams, are orthogonal with respect to the scaling factor. It is also seen from (3.156) that, in contrast to the asymmetric Bessel beams in [87, 88], the L-beams fall into two classes with the even and odd topological charge. The complex amplitudes of beams from two different classes are orthogonal with each other.

In conclusion, the following results were obtained in this section [168]. A new solution of the Helmholtz equation was obtained, describing three-parameter family of the Lommel diffraction-free nonparaxial beams; the complex amplitude of these beams is described by the Lommel functions of two variables, the first of which is complex (equation (3.147)). With an increase in the asymmetry parameter of the Lommel beams, the intensity of the side lobes increases along one Cartesian coordinate and decreases along the other coordinate. The Lommel beams, like any other three-dimensional diffraction-free beams, have annular angular spectrum of the plane waves, which depends only on the polar angular coordinates (equation (3.146)). The Lommel beams have an OAM, which grows linearly with an increase in the mode number n and nonlinearly with an increase in the asymmetry parameter c (equation (3.155)). The functions describing the complex amplitudes of the Lommel beams are orthogonal with respect to the scaling factor α and are not orthogonal with respect to the asymmetry parameter c; according to the mode number n, the beams are orthogonal in the case of different parity.

3.7. Asymmetric Laguerre–Gauss Beams

The Laguerre–Gauss (LG) modes constitute a class of well-studied paraxial light fields. The cross-sectional shape of the intensity of these fields is invariant to propagation in a homogeneous medium and has a radial symmetry. The LG modes have found application in such areas as optical manipulation of micro-objects, quantum optics, and optical communications. Each mode from this class is characterized by two indices – radial and azimuthal, which specify the orbital angular momentum.

Despite the long history and a large amount of research carried out with regard to the LG modes, there still appear publications to study their properties [169–173], formation [174–176], and applications [177–181].

For example, work [169] deals with the propagation of composite vortex beams, representing a coaxial superposition of the LG modes with the same position and waist size. Fields composed of equidistant arrays of solitary or tandem low-intensity spots located on diffractive rings have been generated. In [170], the physical meaning of the radial index of LG modes was investigated, and work [171]] investigates the spatial distribution of the intensity of a sharply focused LG beam, depending on the type of homogeneous (linear and circular) polarization and topological charge. It is shown that the type of polarization has the greatest effect on the longitudinal component of the electric vector of the light field, and the greatest visual difference in the picture of general intensity depending on polarization is observed when using a first-order vortex phase. The nonparaxial propagation of the LG modes in the presence of an aperture was investigated in [172]. It was established that diffraction on the aperture causes a significant distortion of the field in the near zone, but it does not significantly affect the intensity distribution in the far zone (unless the aperture covers a substantial part of the beam). In [173] the properties of light fields with the orbital angular momentum and not having radial symmetry in the presence of a harmonic potential are considered. In [174], a method is proposed for the formation of the LG modes in the cavity of a solid-state laser. Generation of lower-order LG modes with the ability to control the topological charge by means of a solid-state laser was discussed in [175]. In [176], the LG modes with a nonzero radial index are formed using zone spiral phase plates. In [177], it was shown that when using the LG modes instead of the usual Gaussian beam, the Doppler line width in the absorption spectrum of the atoms of rubidium-85 and rubidium-87 decreases, and work [178] considers the use of LG modes to reduce the influence of thermal noise in detectors of gravitational waves. The use of the LG modes for the organization of spin-orbit interaction in ultra-cold atoms is investigated in [179].

The interaction of a LG beam with an atom or a diatomic molecule is studied in [180]. The transfer of the orbital angular momentum between the mass centre and internal motion of a sufficiently cooled atom or molecule has been shown to take place. The three-dimensional off-axis optical trap of dielectric submicron microspheres, created with the help of a single LG beam, is described in [181]. Work [182] is devoted to the application of Laguerre–Gauss modes for quantum communications for a distance of about 3 km in a turbulent atmosphere.

From the above review of the latest literature on the LG modes, it is clear that they not only find new applications, but also are the basis for constructing new types of light fields, which are only theoretically studied. Along with the study of various superpositions of known laser beams, their new types can be obtained by simply shifting the complex amplitude in the Cartesian coordinates by complex distances. For example, the expression for the field of a point source with an imaginary coordinate is reduced to a paraxial Gaussian beam in the case when the imaginary distance from the source to the real space is many times greater than the wavelength of light [183]. In a similar way, asymmetric Bessel diffraction-free modes were obtained, whose intensity distribution in the transverse plane is crescent-shaped [88].

In this section, also using the technique of complex displacement in the Cartesian coordinates, the asymmetric Laguerre–Gauss beams (aLG beams) are theoretically and experimentally investigated. Like the standard LG modes, their intensity distribution in the transverse plane also consists of a finite (as opposed to the Bessel modes) number of light rings, but the intensity distribution on the rings is non-uniform. When propagating in a homogeneous medium, the brightness on the peripheral ring increases. The orbital angular momentum (OAM) and the power of the aLG beam are analytically calculated. As a special case, the aLG beams with zero radial index, having a cross-section of intensity in the form of a crescent, are considered. For such beams, an expression for the coordinates of the intensity maximum is obtained analytically and the rotation of the diffraction pattern during propagation in a homogeneous medium is shown.The aLG beam with the zero radial index was formed using a spatial light modulator and its rotation in space was confirmed experimentally. The possibility of the formation of noncoaxial superpositions of beams with an intensity distribution close to Gaussian, rotating as a whole with propagation in space is shown.

3.7.1. Laguerre–Gauss beams with complex displacement

The complex amplitude of the standard LG beam in the initial plane in the polar coordinates has the form [184]:

$$E_{mn}\left(r,\varphi,z=0\right)=\left(\frac{\sqrt{2}r}{w_0}\right)^n L_m^n\left(\frac{2r^2}{w_0^2}\right)\exp\left(-\frac{r^2}{w_0^2}+in\varphi\right), \qquad (3.157)$$

where $(r,\ \varphi,\ z)$ are the cylindrical coordinates, w_0 is the Gaussian beam waist radius, n is the topological charge of the optical vortex $L_m^n\left(x\right)$, is the associated Laguerre polynomial.

If the beam is shifted by distance x_0 along the x coordinate and by distance y_0 along the y coordinate (displacements x_0 and y_0 can be complex quantities), then in the Cartesian coordinates the beam amplitude is:

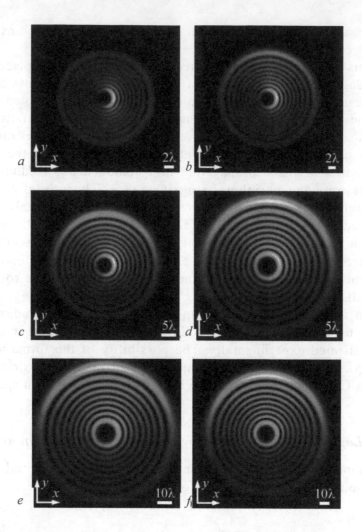

Fig. 3.32. Transverse intensity distribution of an aLG beam in different planes for the following parameters: wavelength $\lambda = 532$ nm, the waist radius $w_0 = 2\lambda$, beam index $(m, n) = (8, 7)$, the transverse displacement $x_0 = 0$, $y_0 = 0.2\lambda i$, the distances along the optical axis are equal to $z = 0\lambda$ (a), 12λ (b), 20λ (c), 40λ (d), 60λ (e), 80λ (f).

$$E_{mn}(x, y, z = 0) = \left(\frac{\sqrt{2}}{w_0}\right)^n \left[(x - x_0) + i(y - y_0)\right]^n \exp\left(-\frac{\rho^2}{w_0^2}\right) L_m^n \left(\frac{2\rho^2}{w_0^2}\right), (3.158)$$

where $\rho^2 = (x - x_0)^2 + (y - y_0)^2$.

When propagating in free space at an arbitrary distance z, the complex amplitude of such a light beam will take the form:

$$E_{mn}(x,y,z) = \frac{w(0)}{w(z)}\left[\frac{\sqrt{2}}{w(z)}\right]^n \left[(x-x_0)+i(y-y_0)\right]^n L_m^n \left[\frac{2\rho^2}{w^2(z)}\right] \times$$

$$\times \exp\left[-\frac{\rho^2}{w^2(z)} + \frac{ik\rho^2}{2R(z)} - i(n+2m+1)\zeta(z)\right], \tag{3.159}$$

where

$$w(z) = w_0\sqrt{1+\left(\frac{z}{z_R}\right)^2},$$

$$R(z) = z\left[1+\left(\frac{z_R}{z}\right)^2\right], \tag{3.160}$$

$$\zeta(z) = \arctan\left(\frac{z}{z_R}\right),$$

$z_R = kw_0^2/2$ is the Rayleigh distance, $k = 2\pi/\lambda$ is the wave number of light with wavelength λ.

If the displacements x_0 and y_0 are not real values, the values ρ^2, $w(z)R(z)$ does not already have the physical meaning of that for real displacements x_0 and y_0, i.e., it is not the distance to the optical axis, not the width of the beam, and not the radius of curvature of the wave front. In addition, unlike the standard LG modes, the cross-sectional intensity of such a beam does not have radial symmetry. Figure 3.32 shows the results of the calculation of the transverse intensity of the beam (3.159) at different distances for the following parameters: wavelength $\lambda = 532$ nm, waist radius $w_0 = 2\lambda$, beam index $(m, n) = (8, 7)$, the transverse displacement $x_0 = 0$, $y_0 = 0.2\lambda i$, the distance along the optical axis are equal to $z = 0\lambda$ (Fig. 3.32 *a*), 12λ (Fig. 3.32 *b*), 20λ (Fig. 3.32 *c*), 40λ (Fig. 3.32 *d*), 60λ (Fig. 3.32 *e*), 80λ (Fig. 3.32 *f*). The size of the computational domain is $2R$, where $R = 15\lambda$ (Fig. 3.32 *a*), 20λ (Fig. 3.32 *b*), 25λ (Fig. 3.32 *c*), 35λ (Fig. 3.32 *d*), 50λ (Fig. 3.32 *e*), 75λ (Fig. 3.32 *f*). For the given parameters, the Rayleigh distance is equal to $z_R = 4\pi\lambda$.

From Fig. 3.32 it is seen that with the propagation of the beam, the first crescent-shaped ring becomes almost a ring, although the peripheral rings are not fully restored. It is also seen that during the propagation of the aLG beam, the energy is redistributed from the crescent to the peripheral rings.

3.7.2. Power of the shifted Laguerre–Gauss beam

The power of an arbitrary paraxial light beam can be calculated through both the complex amplitude E and the angular spectrum of plane waves A:

$$W = \iint_{\mathbb{R}^2} E^* E \, dx \, dy = \lambda^2 \iint_{\mathbb{R}^2} A^* A \, d\alpha \, d\beta, \tag{3.161}$$

where

$$A(\alpha,\beta) = \lambda^{-2} \iint_{\mathbb{R}^2} E(x,y,0) \exp\left[-ik(\alpha x + \beta y)\right] dx \, dy. \tag{3.162}$$

For an aLG beam, it is more convenient to calculate the power in the spectral plane. First, we obtain the expression for the angular spectrum of plane waves of an unshifted beam:

$$A(\rho,\theta) = \frac{1}{\lambda^2}\left(\frac{\sqrt{2}}{w_0}\right)^n \int_0^\infty r^n \exp\left(-\frac{r^2}{w_0^2}\right) L_m^n\left(\frac{2r^2}{w_0^2}\right) \times$$

$$\times \left\{ \int_0^{2\pi} \exp\left[in\varphi - ik\rho r \cos(\varphi-\theta) \right] d\varphi \right\} r \, dr = \tag{3.163}$$

$$= (-i)^n \frac{k^2}{2\pi}\left(\frac{\sqrt{2}}{w_0}\right)^n \exp(in\theta) \int_0^\infty r^{n+1} \exp\left(-\frac{r^2}{w_0^2}\right) L_m^n\left(\frac{2r^2}{w_0^2}\right) J_n(k\rho r) \, dr,$$

where (r, φ) and (ρ, θ) are the polar coordinates in the initial plane and in the Fourier plane, respectively (ρ is the dimensionless coordinate).

The reference integral ([112], expression 7.421.4) is known:

$$\int_0^\infty x^{v+1} \exp\left(-\beta x^2\right) L_n^v\left(\alpha x^2\right) J_v(xy) \, dx =$$

$$= \frac{(\beta-\alpha)^n y^v}{2^{v+1} \beta^{v+n+1}} \exp\left(-\frac{y^2}{4\beta}\right) L_n^v\left[\frac{\alpha y^2}{4\beta(\alpha-\beta)}\right]. \tag{3.164}$$

Using it, we obtain the expression for the angular spectrum of plane waves of the LG beam:

$$A(\rho,\theta) = C_0 \rho^n L_m^n\left[\frac{(kw_0\rho)^2}{2}\right] \exp\left[-\frac{(kw_0\rho)^2}{4} + in\theta\right], \tag{3.165}$$

where

$$C_0 = (-i)^n (-1)^m \frac{(kw_0)^{n+2}}{2^{2+n/2} \pi}. \tag{3.166}$$

For a shifted beam (3.158), the expression for the angular spectrum of plane waves is:

$$A(\rho, \theta) = C_0 \rho^n \exp\left[-\frac{(kw_0 \rho)^2}{4} + in\theta \right] \times$$

$$\times L_m^n \left[\frac{(kw_0 \rho)^2}{2} \right] \exp\left[-ik\rho(x_0 \cos\theta + y_0 \sin\theta) \right]. \tag{3.167}$$

Using expression (3.167), we find the power of the aIG beam:

$$W = 2\pi \lambda^2 |C_0|^2 \int_0^\infty \rho^{2n+1} \exp\left[-\frac{(kw_0 \rho)^2}{2} \right] \times$$

$$\times \left\{ L_m^n \left[\frac{(kw_0 \rho)^2}{2} \right] \right\}^2 J_0(2ikD_0 \rho) d\rho, \tag{3.168}$$

where $D_0 = [(\mathrm{Im} x_0)^2 + (\mathrm{Im} y_0)^2]^{1/2}$.

The integral (3.168) can be calculated using the reference integral (expression 2.9.12.14 in [65], corrected after numerical verification):

$$\int_0^\infty \overline{uuuuuu}^{-cx} \,_{\gamma+\delta}\left(\sqrt{} \right) \,_{\mu}^{\gamma}() \,_{\nu}^{\delta}() =$$

$$= \frac{(-1)^{\mu+\nu}}{c^{\gamma+\delta+1}} \left(\frac{b}{2} \right)^{\gamma+\delta} \exp\left(-\frac{b^2}{4c} \right) L_\nu^{\gamma+\mu-\nu}\left(\frac{b^2}{4c} \right) L_\mu^{\delta-\mu+\nu}\left(\frac{b^2}{4c} \right), \tag{3.169}$$

where $\mathrm{Re} > 0$, $\mathrm{Re}(\gamma + \lambda) > -1$, $|\arg b| < \pi$.

Replacing in (3.169) integration variable $x \to x^2$ and setting $\mu = m$, $\gamma = n$, $\delta = -n$, $\nu = m+n$, $c = (kw_0)^2/2$, $b = 2ikD_0$. Then, in view of the identity

$$L_\mu^{-\sigma}(x) \equiv \left[(\mu - \sigma)!/\mu! \right](-x)^\sigma L_{\mu-\sigma}^\sigma(x)$$

and replacing the integration variable $x \to x^2$, we obtain the expression for the beam power:

$$W = \frac{\pi w_0^2}{2} \frac{(m+n)!}{m!} \exp\left(\frac{2D_0^2}{w_0^2}\right) L_{m+n}\left(-\frac{2D_0^2}{w_0^2}\right) L_m\left(-\frac{2D_0^2}{w_0^2}\right). \qquad (3.170)$$

Despite the proportionality of the power of the beam to the Laguerre polynomials, the power cannot be negative or equal to zero. This follows from the fact that $2(D_0/w_0)^2 \geq 0$, and Laguerre polynomials in the domain of non-positive values are always positive:

$$L_m(-\xi) = \sum_{k=0}^{m} \frac{(-1)^k}{k!} C_m^k (-\xi)^k = \sum_{k=0}^{m} C_m^k \frac{\xi^k}{k!} = 1 + \underbrace{\sum_{k=0}^{m-1} C_m^k \frac{\xi^k}{k!}}_{\geq 0} \geq 1. \qquad (3.171)$$

From (3.171) and from the presence of the factor $\exp[2(D_0/w_0)^2]$ in (3.170) it follows that a shift by a complex value always leads to an increase in the beam power.

In the particular case, when the beam is shifted by real distances, the parameter D_0 becomes zero and the power takes the value $[\pi w_0^2/2][(m+n)!/m!]$, which coincides with the expression for the power of the LG beams from [184] with the accuracy to a constant.

There is no physical sense in increasing the power with increasing asymmetry of the aLG beam. But the normalized OAM in the next section will be calculated using the expression (3.170).

3.7.3. Orbital angular momentum of the shifted Laguerre–Gauss beam

We now obtain an expression for the projection of the aLG beam on the optical axis (the remaining projections for the paraxial beams are zero) This is also convenient to do through the angular spectrum of plane waves:

$$J_z = -i\lambda^2 \iint_{\mathbb{R}^2} A^* \frac{\partial A}{\partial \theta} \rho d\rho d\theta. \qquad (3.172)$$

Substituting (3.167) into (3.172), we get:

$$J_z = -i\lambda^2 |C_0|^2 \iint_{\mathbb{R}^2} \rho^{2n} \exp\left[-\frac{(kw\rho)^2}{2}\right] \left\{ L_m^n\left[\frac{(kw\rho)^2}{2}\right] \right\}^2 \times$$

$$\times \exp\left[-in\theta + ik\rho\left(x_0^* \cos\theta + y_0^* \sin\theta\right)\right]\left[in + ik\rho\left(x_0 \sin\theta - y_0 \cos\theta\right)\right] \times$$

$$\times \exp\left[in\theta - ik\rho\left(x_0 \cos\theta + y_0 \sin\theta\right)\right]\rho d\rho d\theta. \qquad (3.173)$$

Let us separate the term proportional to the power:

$$J_z = nW + k\lambda^2 |C_0|^2 \int_0^\infty \rho^{2n+2} \exp\left[-\frac{(kw\rho)^2}{2}\right] \left\{L_m^n\left[\frac{(kw\rho)^2}{2}\right]\right\}^2 d\rho \times$$

$$\times \int_0^{2\pi} (x_0 \sin\theta - y_0 \cos\theta) \exp\{2k\rho[(\operatorname{Im} x_0)\cos\theta + (\operatorname{Im} y_0)\sin\theta]\} d\theta.$$

(3.174)

The internal integral is expressed in terms of the Bessel functions [154]. Then (3.174) will take the form:

$$J_z = nW - 2\pi k\lambda^2 |C_0|^2 \int_0^\infty \rho^{2n+2} \exp\left[-\frac{(kw\rho)^2}{2}\right] \left\{L_m^n\left[\frac{(kw\rho)^2}{2}\right]\right\}^2 \times$$

$$\times \left\{(ix_0 + y_0)\left[\frac{(\operatorname{Im} x_0) - (\operatorname{Im} y_0)}{2iD_0}\right] + (ix_0 - y_0)\left[\frac{(\operatorname{Im} x_0) - (\operatorname{Im} y_0)}{2iD_0}\right]^{-1}\right\} \times$$

$$\times iI_1(2kD_0\rho) d\rho = nW + 4\pi^2\lambda|C_0|^2 \frac{\operatorname{Im}(x_0^* y_0)}{D_0} \times$$

$$\times \int_0^\infty \rho^{2n+2} \exp\left[-\frac{(kw\rho)^2}{2}\right] \left\{L_m^n\left[\frac{(kw\rho)^2}{2}\right]\right\}^2 I_1(2kD_0\rho) d\rho.$$

(3.175)

This integral can be calculated as the integral for the beam power, if we represent it as a derivative with respect to D_0:

$$J_z = nW - 4\pi^2\lambda|C_0|^2 \frac{\operatorname{Im}(x_0^* y_0)}{2kD_0} \frac{\partial}{\partial D_0}\left(\int_0^\infty \rho^{2n+1} \exp\left[-\frac{(kw\rho)^2}{2}\right] \times \right.$$

$$\left. \times \left\{L_m^n\left[\frac{(kw\rho)^2}{2}\right]\right\}^2 I_0(2kD_0\rho) d\rho\right).$$

(3.176)

The integral in (3.176) is taken and coincides with the integral for calculating the power:

$$J_z = nW - 4\pi^2\lambda|C_0|^2 \frac{\mathrm{Im}(x_0^*y_0)}{2kD_0} \frac{\partial}{\partial D_0} =$$

$$= \left\{ \frac{2^n}{(kw_0)^{2n+2}} \frac{(m+n)!}{m!} \exp\left(\frac{2D_0^2}{w_0^2}\right) L_{m+n}\left(-\frac{2D_0^2}{w_0^2}\right) L_m\left(-\frac{2D_0^2}{w_0^2}\right) \right\} =$$

$$= nW - \frac{\pi w_0^2}{4} \frac{(m+n)!}{m!} \frac{\mathrm{Im}(x_0^*y_0)}{D_0} \frac{\partial}{\partial D_0} \times$$

$$\times \left\{ \exp\left(\frac{2D_0^2}{w_0^2}\right) L_{m+n}\left(-\frac{2D_0^2}{w_0^2}\right) L_m\left(-\frac{2D_0^2}{w_0^2}\right) \right\}. \tag{3.177}$$

Calculating the derivative, we obtain the expression for the normalized OAM:

$$\frac{J_z}{W} = n + \frac{2\,\mathrm{Im}(x_0^*y_0)}{w_0^2} \left[\frac{L_m^1\left(-\frac{2D_0^2}{w_0^2}\right)}{L_m\left(-\frac{2D_0^2}{w_0^2}\right)} + \frac{L_{m+n}^1\left(-\frac{2D_0^2}{w_0^2}\right)}{L_{m+n}\left(-\frac{2D_0^2}{w_0^2}\right)} - 1 \right]. \tag{3.178}$$

From (3.178) it can be seen that the normalized OAM (OAM per photon) does not depend on the wavelength, but is completely determined by the ratio of displacements to the waist radius, i.e. values x_0/w_0 and y_0/w_0.

It can be shown that an increase or decrease in the nirmalized OAM is completely determined by the sign of the value $\mathrm{Im}(x_0^*y_0)$, since the expression in square brackets is always greater than or equal to 1.

The normalized OAM depends at once on several parameters – on both indices of the Laguerre polynomials and on the real and imaginary displacements along both Cartesian coordinates. Let us consider a particular case when the value $x_0^*y_0$ is purely imaginary, and the complex displacements are equal in magnitude. Let $x_0 = aw_0\exp(i\upsilon)$, $y_0 = ix_0$, where a is the normalized modulus of complex displacement (i.e., $a = |x_0/w_0|$). Then the expression for the OAM takes the form:

$$\frac{J_z}{W} = n - \xi\left[\frac{L_m^1(\xi)}{L_m(\xi)} + \frac{L_{m+n}^1(\xi)}{L_{m+n}(\xi)} - 1 \right], \tag{3.179}$$

where $\xi = -2a^2$. The parameter a can be called the asymmetry parameter of an aLG beam. From (3.179) it is clear that for

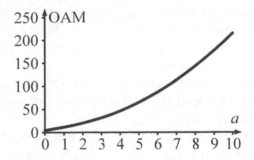

Fig. 3.33. Dependence of the normalized OAM on the normalized asymmetry parameter a.

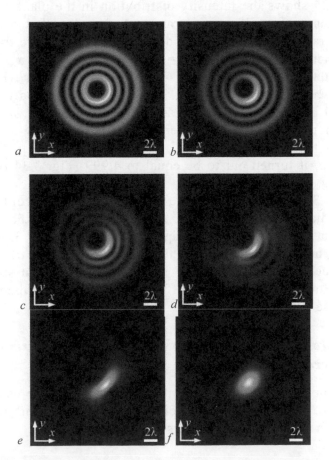

Fig. 3.34. Intensity distributions in the plane $z = 0$ for aLG beams at the following parameters: wavelength $\lambda = 532$ nm, the waist radius $w = 2\lambda$, beam index $(m,n) = (3, 5)$, the transverse displacement $x_0 = 0.01wi$ and $y_0 = 0.01wi$ (*a*), $x_0 = 0.05wi$ and $y_0 = 0.05wi$ (*b*), $x_0 = 0.1wi$ and $y_0 = 0.1wi$ (*c*), $x_0 = 0.2wi$ and $y_0 = 0.2wi$ (*d*), $x_0 = 0.5wi$ and $y_0 = 0.5wi$ (*d*), $x_0 = 2wi$ and $y_0 = 2wi$ (*f*).

$|\xi| \gg 1$ the normalized OAM is approximately $J_z/W \approx n - \xi$, i.e. depends parabolically on the asymmetry parameter a, which is confirmed by the graph in Fig. 3.33, calculated by formula (3.179) for $m = 3$ and $n = 5$.

From Fig. 3.33 it can be seen that, in contrast to the ordinary LG modes, the OAM of the aLG beams varies continuously, running integer and fractional values.

From (3.178) it also follows that at $x_0^* y_0 \in \mathbb{R}$ with the normalized OAM coincides with the topological charge n, as in the case of radially symmetric optical vortices, although the beam shape does not have radial symmetry (if x_0 and y_0 are purely imaginary). So, Figure 3.34 shows the intensity distribution in the plane $z = 0$ for aLG beams at the following parameters: wavelength $\lambda = 532$ nm, the waist radius $w = 2\lambda$, beam index $(m, n) = (3, 5)$, the transverse displacement $x_0 = 0.01wi$ and $y_0 = 0.01wi$ (Fig. 3.34 *a*), $x_0 = 0.05wi$ and $y_0 = 0.05wi$ (Fig. 3.34 *b*), $x_0 = 0.1wi$ and $y_0 = 0.1wi$ (Fig. 3.34 *c*), $x_0 = 0.2wi$ and $y_0 = 0.2wi$ (Fig. 3.34 *d*), $x_0 = 0.5wi$ and $y_0 = 0.5wi$ (Fig. 3.34 *e*), $x_0 = 2wi$ and $y_0 = 2wi$ (Fig. 3.34 *f*). The size of the computational domain is $2R$, where $R = 10\lambda$. According to (3.178), the OAM for all graphs in Fig. 3.34 must be equal to 5. When calculating, it turned out to be equal to 4.999 (Fig. 3.34 *a–d*) and 4.998 (Fig. 3.34 *e, f*).

Figure 3.34 shows that these beams have a different form – almost radially symmetrical light rings in Fig. 3.34 *a*, the light arc (a crescent), surrounded by peripheral rings, in Figs. 3.34 *b* and 3.34 *c*, a broken crescent peripheral rings (Fig. 3.34 *d*), without crescent peripheral rings (Fig. 3.34 *e*), an elliptic light spot (Fig. 3.34 *f*). However, despite the different type, the normalized OAM of all these beams is the same.

3.7.4. Paraxial Laguerre–Gaussian beams in the form of a rotating crescent

Expression (3.159) is significantly simplified in the case when $m = 0$, i.e. the diffraction pattern has only one light ring. The intensity in this case is as follows:

$$I_{0n}(x,y,z) = |E|^2 = \frac{w^2(0)}{w^2(z)} \left[\frac{\sqrt{2}}{w^2(z)} \right]^n \times$$

$$\times \exp\left[2\frac{(\operatorname{Im} x_0)^2 + (\operatorname{Im} y_0)^2}{w^2(z)} \right] \times \left[(u + \operatorname{Im} y_0)^2 + (v - \operatorname{Im} x_0)^2 \right]^n \times \qquad (3.180)$$

$$\times \exp\left\{ -\frac{2(u^2 + v^2)}{w^2(z)} + \frac{2k\left[(\operatorname{Im} x_0)u + (\operatorname{Im} y_0)v \right]}{R(z)} \right\}.$$

where $u = x - \operatorname{Re} x_0$, $v = y - \operatorname{Re} y_0$. Further, for brevity, we introduce the notation $D_0 = [(\operatorname{Im} x_0)^2 + (\operatorname{Im} y_0)^2]^{1/2}$.

The intensity cannot be negative, therefore the intensity zeros are its minima. From (3.180), it can be seen that with the propagation of an aLG beam in a uniform space for $n > 0$, there is a minimum of intensity at the centre of the beam (point of phase singularity) with coordinates $(x_{min}, y_{min}) = (\operatorname{Re} x_0 - \operatorname{Im} y_0, \operatorname{Re} y_0 + \operatorname{Im} x_0)$. Equating to zero the partial derivatives of intensity (3.180) with respect to the Cartesian coordinates, it can be shown that for $n > 0$, the position of the intensity maximum depends on the distance covered z and rotates relative to the minimum:

$$\frac{y_{max} - \operatorname{Re} y_0 - \operatorname{Im} x_0}{x_{max} - \operatorname{Re} x_0 + \operatorname{Im} y_0} = \frac{(\operatorname{Im} y_0)z - (\operatorname{Im} x_0)z_R}{(\operatorname{Im} x_0)z + (\operatorname{Im} y_0)z_R}, \qquad (3.181)$$

where (x_{max}, y_{max}) are the coordinates of the point with the maximum intensity.

We introduce a new coordinate system shifted and rotated relative to the original system by an angle depending on the distance travelled z:

$$\begin{pmatrix} \xi \\ \eta \end{pmatrix} = \begin{pmatrix} \cos\varphi & \sin\varphi \\ -\sin\varphi & \cos\varphi \end{pmatrix} \begin{pmatrix} x - \operatorname{Re} x_0 + \operatorname{Im} y_0 \\ y - \operatorname{Re} y_0 - \operatorname{Im} x_0 \end{pmatrix}, \qquad (3.182)$$

Where

$$\varphi = \arctan\left[\frac{(\operatorname{Im} y_0)z - (\operatorname{Im} x_0)z_R}{(\operatorname{Im} x_0)z + (\operatorname{Im} y_0)z_R} \right], \quad \cos\varphi = \frac{(\operatorname{Im} x_0)z + (\operatorname{Im} y_0)z_R}{D_0\sqrt{z^2 + z_R^2}},$$

$$\sin\varphi = \frac{(\operatorname{Im} y_0)z - (\operatorname{Im} x_0)z_R}{D_0\sqrt{z^2 + z_R^2}}. \qquad (3.183)$$

In the new coordinate system (ξ, η), the intensity (3.180) takes the form:

$$I_{0n}(\xi,\eta,z) = |E|^2 = \frac{w^2(0)}{w^2(z)}\left[\frac{\sqrt{2}}{w^2(z)}\right]^n \times$$

$$\times \left(\xi^2+\eta^2\right)^n \exp\left[-\frac{2\left(\xi^2+\eta^2\right)}{w^2(z)} + \frac{2kD_0}{\sqrt{z^2+z_R^2}}\xi\right]. \tag{3.184}$$

It can be shown that such a function has three stationary points. The first of them is the minimum point $\xi = 0$, $\eta = 0$, corresponding to the above-mentioned point (x_{min}, y_{min}). The second – the maximum point with coordinates

$$\begin{cases} \xi_{max} = \dfrac{1}{2}\dfrac{w(z)}{w(0)}\left(D_0 + \sqrt{D_0^2 + 2nw_0^2}\right), \\ \eta_{max} = 0, \end{cases} \tag{3.185}$$

and the third is a saddle point with the coordinates

$$\begin{cases} \xi_{saddle} = \dfrac{1}{2}\dfrac{w(z)}{w(0)}\left(D_0 - \sqrt{D_0^2 + 2nw_0^2}\right), \\ \eta_{saddle} = 0, \end{cases} \tag{3.186}$$

which is the maximum point in the variable ξ and the minimum point in the variable η. The maximum intensity in the variable ξ means that the point lies on the light ring, and the minimum in the variable η means that this is the point with the minimum intensity on the ring.

In the original coordinate system, the maximum intensity point has the following coordinates:

$$\begin{cases} x_{max} = \operatorname{Re} x_0 - \operatorname{Im} y_0 + \\ \quad + \dfrac{(\operatorname{Im} x_0)z + (\operatorname{Im} y_0)z_R}{2z_R}\left(1 + \sqrt{1 + \dfrac{2nw_0^2}{D_0^2}}\right), \\ y_{max} = \operatorname{Re} y_0 + \operatorname{Im} x_0 + \\ \quad + \dfrac{(\operatorname{Im} y_0)z - (\operatorname{Im} x_0)z_R}{2z_R}\left(1 + \sqrt{1 + \dfrac{2nw_0^2}{D_0^2}}\right). \end{cases} \tag{3.187}$$

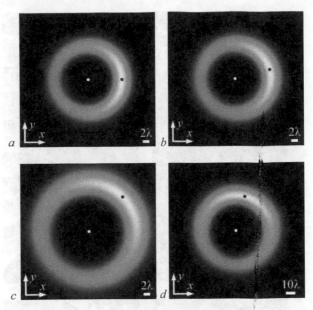

Fig. 3.35. The transverse distribution of the beam intensity (3.159) with zero radial index $m = 0$ at different distances for the following parameters: wavelength $\lambda = 532$ nm, waist radius $w_0 = 5\lambda$, topological charge $n = 8$, the transverse displacement $x_0 = 0.25\lambda = w_0/20$, and $y_0 = 0.25\lambda i = iw_0/20$, the distances along the optical axis are equal $z = 0$ (*a*), $z_R \tan(\pi/12)$ (*b*), $z_R \tan(\pi/4) = z_R$ (*c*), $z_R \tan(5\pi/12)$ (*d*). The size of the computational domain is $2R$, where $R = 20\lambda$ (*a*), 20λ (*b*), 20λ (*c*), 75λ (*d*). The black point indicates the position of the maximum intensity, calculated by the formula (3.187).

From (3.187) it follows that the point with the maximum intensity (x_{max}, y_{max}) rotates around the point (x_{min}, y_{min}), turning through an angle α_0 at a distance

$$z = z_R \tan(\alpha_0). \tag{3.188}$$

Figure 3.35 shows the results of calculation of the transverse beam intensity (3.159) with the zero radial index $m = 0$ in different planes for the following parameters: wavelength $\lambda = 532$ nm, the waist radius $w_0 = 5\lambda$ (i.e., the Rayleigh distance is equal to $z_R = 25\pi\lambda$), topological charge $n = 8$, the transverse displacement $x_0 = 0.25\lambda = w_0/20$, and $y_0 = 0.25\lambda i = iw_0/20$, the distance along the optical axis are equal to $z = 0$ (Fig. 3.35 *a*), $z_R \tan(\pi/12)$ (Fig. 3.35 *b*), $z_R \tan(\pi/4) = z_R$ (Fig. 3.35 *c*), $z_R \tan(5\pi/12)$ (Fig. 3.35 *d*). The centre of the beam is at the origin: $(x_{min}, y_{min}) = (0, 0)$. The size of the computational domain is $2R$, where $R = 20\lambda$ (Fig. 3.35 *a*), 20λ (Fig. 3.35 *b*), 20λ

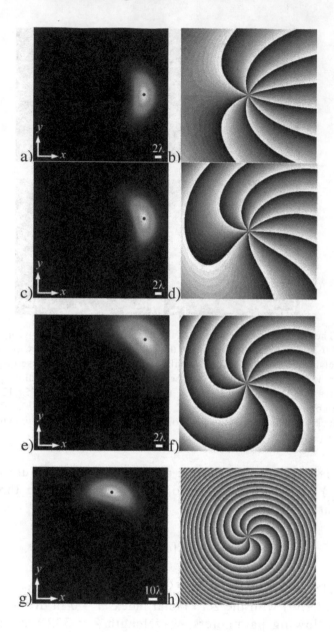

Fig. 3.36. Transverse beam intensity distribution (3.159) with zero radial index $m = 0$ at different distances z for the following parameters: wavelength $\lambda = 532$ nm, the waist radius $w_0 = 5\lambda$, topological charge $n = 8$, the transverse displacement $x_0 = 5\lambda = w_0$ and $y_0 = 5\lambda i = iw_0$, the distances along the optical axis are $z = 0$ (a), $z_R \tan(\pi/12)$ (b), $z_R \tan(\pi/4) = z_R$ (c), $z_R \tan(5\pi/12)$ (d). The size of the computational domain is $2R$, where $R = 20\lambda$ (a), 20λ (b), 20λ (c), 75λ (d). The black point marks the position of the intensity maximum, calculated by formula (3.187).

(Fig. 3.35 *c*), 75λ (Fig. 3.35 *d*). The black point marks the position of the intensity maximum, calculated by formula (3.187). The white point marks the minimum intensity points or phase singularity points.

From (3.184)–(3.186) it follows that the ratio of the maximum intensity on the light ring to the minimum (that is, at the saddle point) is determined by the following expression:

$$
\frac{I_{0n,\max}}{I_{0n,\text{saddle}}} = \left(\frac{\sqrt{D_0^2 + 2nw_0^2} + D_0}{\sqrt{D_0^2 + 2nw_0^2} - D_0} \right)^{2n} \exp\left(\frac{2D_0}{w_0^2} \sqrt{D_0^2 + 2nw_0^2} \right). \tag{3.189}
$$

From (3.189) it follows that the asymmetry of the beam grows exponentially with increasing displacement D_0. This means that by increasing the imaginary displacements Im (x_0) and Im (y_0) there is a 'break' of the light rings and the intensity distribution has the form of a light spot shifted from the centre and rotating during propagation relative to the phase singularity point $(x_{\min}, y_{\min})$. The word 'break' is in quotation marks, since there can be no zero intensity on the light ring (the denominator in (3.189) never vanishes with $n > 0$).

Figure 3.36 shows the results of calculating the transverse intensity and phase of the same beam as in Fig. 3.35, but with large displacements $x_0 = 5\lambda = w_0$ and $y_0 = 5\lambda i = iw_0$. All other parameters are the same as in Fig. 3.35. The black point marks the position of the intensity maximum, calculated by formula (3.187).

From Fig. 3.36 it can be seen that the elongated light spot rotates by $\pi/12$, $\pi/4$, $5\pi/12$, respectively, while the centre of the phase singularity remains at the origin.

The asymmetric LG beam at $m = 0$ is similar to the asymmetric Bessel–Gauss beam (aBG beam) [124]. Both beams have the same rotation speed, determined by formula (3.188). However, an aBG beam is more difficult to analytically describe. The formulas for the coordinates of intensity zeros were obtained in [124], but there are no formulas for determining the coordinates of the intensity maxima, like (3.187). The OAM of the aBG beam is described in terms of the series of modified Bessel functions, in contrast to the closed expression for the aLG beams (3.178). In addition, the OAM of the aLG beam is in a parabolic dependence on the asymmetry parameter, and the OAM of the aBG beam in a dependence close to linear [124].

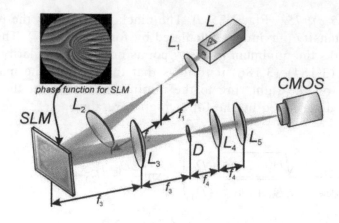

Fig. 3.37. Experimental setup scheme for the formation of aLG beams: L – solid-state laser ($\lambda = 532$ nm), L_1, L_2, L_3, L_4 and L_5 – lenses with focal lengths ($f_1 = 250$ mm, $f_2 = 500$ mm, $f_3 = 350$ mm, $f_4 = 150$ mm, $f_5 = 280$ mm), SLM – spatial light modulator PLUTO VIS (resolution 1920 × 1080 pixels, pixel size 8 microns), D – aperture acting as a spatial filter, CMOS - video camera LOMO TC-1000 (resolution 3664 × 2740 pixels, pixel size 1.67 microns).

3.7.5. Experimental formation of an asymmetric Laguerre–Gauss beam

The optical scheme of the experiment is shown in Fig. 3.37. The output beam of the solid-state laser L ($\lambda = 532$ nm) was expanded using a system consisting of lenses L_1 ($f_1 = 250$ mm) and L_2($f_2 = 500$ mm). After expansion, the radius of the laser beam was about 1.1 mm. The expanded laser beam was directed to the display of a spatial light modulator SLM (PLUTO VIS, resolution 1920 × 1080 pixels, pixel size 8 μm). The modulator displayed the phase function, which is a superposition of the encoded phase function of the original element forming the aLG beam, and the linear phase mask (shown in the inset in Fig. 3.37). This was done in order to separate the first diffraction order and the zero diffraction order in which the unmodulated radiation is reflected. Reflected in the first order laser beam after passing a system of lenses L_3($f_3 = 350$ mm) and L_4($f_4 = 150$ mm) was directed to the lens L_5($f_5 = 280$ mm), which focused the aLG beam on the CMOS matrix of the LOMO TC-1000 camera (resolution 3664 × 2740 pixels, pixel size 1.67 μm). The CMOS camera is mounted on the optical rail and moves along it to capture the intensity distribution at different distances from the lens L_5. The diaphragm D plays the role of a spatial filter and blocks the zero diffraction order.

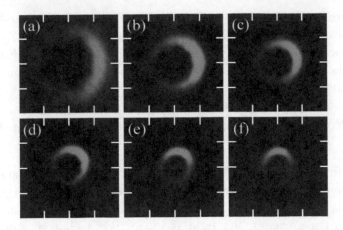

Fig. 3.38. Intensity distributions of the aLG beam, obtained at different distances from the lens plane L_5: (*a*) 0 mm, (*b*) 100 mm, (*c*) 150 mm, (*d*) 200 mm, (*e*) 250 mm, (*f*) 280 mm. The scale size is 500 μm. Parameters: $w_0 = 1$ mm, $n = 8$, $x_0 = 0.2\ w_0$, $y_0 = 0.2\ w_0 i$.

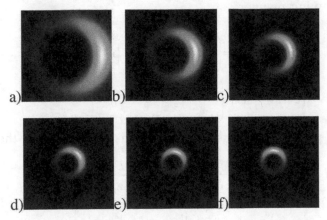

Fig. 3.39. The intensity distribution of the aLG beam, calculated by the formula (3.159) (with the lenses taken into account) with the same parameters as in Fig. 3.38.

Figure 3.38 shows the intensity distributions obtained at different distances from the lens L_5. To produce an image at a distance of 0 mm, the lens L_5 was temporarily removed from the scheme. From the presented images it can be seen that as the crescent-shaped beam is propagates from the plane of the lens, it turns around its axis. The beam formed in the focus of the lens L_5 is rotated 90 degrees relative to the original beam formed in the plane of the lens L_5. In this case, due to the focusing, the transverse dimensions of the formed beams also decrease.

Figure 3.39 shows the intensity distribution calculated by the formula (3.159) (but with the lenses taken into account) with the same parameters as the parameters of the experiment. From Fig. 3.39 it is seen that the calculated and experimental distributions are qualitatively the same.

3.7.6. Rotating superposition of asymmetric Laguerre–Gauss beams

With a further increase in the imaginary displacements x_0 and y_0, the beam asymmetry increases so much that the intensity on the ring is concentrated near the maximum point defined by expression (3.187). Instead of a crescent, the field is more like a Gaussian beam, but shifted from the origin and rotated relative to it when propagating in space by an angle of $\pi/2$.

This follows from the fact that, according to (3.184), in the initial plane in the rotated coordinates (3.182) the intensity is distributed in the following manner:

$$I_{0n}\left(\xi,\eta,z=0\right)=\left(\frac{\sqrt{2}}{w_0^2}\right)^n \exp\left(\frac{2D_0^2}{w_0^2}\right)\times$$

$$\times\left(\xi^2+\eta^2\right)^n \exp\left\{-\frac{2}{w_0^2}\left[\left(\xi-D_0\right)^2+\eta^2\right]\right\}. \tag{3.190}$$

In other planes at a distance z, the intensity (3.190) has the form:

$$I_{0n}\left(\xi,\eta,z\right)=\frac{w^2(0)}{w^2(z)}\left[\frac{\sqrt{2}}{w^2(z)}\right]^n \exp\left(\frac{2D_0^2}{w_0^2}\right)\times$$

$$\times\left(\xi^2+\eta^2\right)^n \exp\left(-\frac{2}{w^2(z)}\left\{\left[\xi-D(z)\right]^2+\eta^2\right\}\right). \tag{3.191}$$

where $D(z) = D_0[1+(z/z_R)^2]^{1/2} = w(z)/w(0)$.

Thus, for $D_0 \gg w_0$, the power-law component $(\xi^2+\eta^2)^n$ has little effect on the form of the intensity distribution, and the remaining exponent in (3.191) corresponds to the intensity of a Gaussian beam with a waist radius $w(z)$ shifted by a distance $D(z) \approx \xi_{max}$ from the origin (the point with the phase singularity). This means that if we consider the superposition of beams (3.190) with different offsets D_0, they will look like non-coaxial Gaussian beams, which, when propagated, turn through the same angle defined by (3.188).

With that, according to (3.191), when the distance from the initial plane z increases the displacement of the intensity maxima from the optical axis increases by $[1+(z/z_R)^2]^{1/2}$ times, i.e. proportional to the expansion of the Gaussian beam. This means that the beams in the superposition almost do not interfere with each other and the diffraction pattern does not change its appearance during propagation.

So, in Fig. 3.40 shows the propagation of two aLG beams, the complex amplitude of which has the form:

$$E(x,y,z) = \frac{w(0)}{w(z)}\left[\frac{\sqrt{2}}{w(z)}\right]^n \sum_{j=1}^{N} C_j\left[(x-x_{0j})+i(y-y_{0j})\right]^n \times$$

$$\times \exp\left[-\frac{\rho_j^2}{w^2(z)} + \frac{ik\rho_j^2}{2R(z)} - i(n+1)\zeta(z)\right],$$

(3.192)

where $N = 2$ for two beams and $N = 3$ for three beams, (x_{0j}, y_{0j}) are the displacements of the j-th beam, and $\rho_j = [(x - x_{0j})^2+(y - y_{0j})^2]^{1/2}$. The coefficients C_j were selected so that the maximum intensity of all beams was the same. Other calculation parameters were selected as follows: wavelength $\lambda = 532$ nm, the waist radius $w_0 = 5\lambda$ (Rayleigh

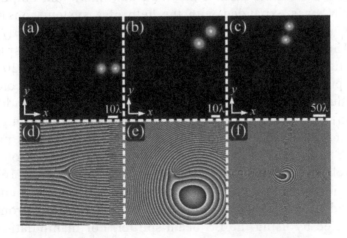

Fig. 3.40. The intensity (a–c) and phase (d–f) distributions of aLG beams at the following values of parameters: wavelength $\lambda = 532$ nm, the radius of the beam waist $w_0 = 5\lambda$, the index of the beam $(m, n) = (0, 8)$, transverse displacement of the first beam $x_0 = 5\ w_0$, $y_0 = 5\ w_0 i$, transverse displacements of the second beam $x_0 = 8w_0$, $y_0 = 8w_0 i$, distances along the optical axis $z = 0$ (a, d), $z = z_R \tan(\pi/4)$ (b, e), $z = z_R \tan(5\pi/12)$ (c, f). The size of the computational domain is $2R$, where $R = 50\lambda$ (a, d), 60λ (b, e), 200λ (c, f).

distance $z_R = 25\pi\lambda$), the beam index $(m, n) = (0, 8)$, transverse displacement of the first beam $x_0 = 5w_0, y_0 = 5w_0 i$, the transverse displacement of the second beam $x_0 = 8w_0$, $y_0 = 8w_0 i$, the distances along the optical axis are equal to $z = 0$ (Fig. 3.40 *a, d*), $z = z_R \tan(\pi/4)$ (Fig. 3.40 (*b, e*)), $z = z_R \tan(5\pi/12)$ (Fig. 3.40 *c, f*). The size of the calculated area is $2R$, where $R = 50\lambda$ (Fig. 3.40 *a, d*), 60λ (Fig. 3.40 *b, e*), 200λ (Fig. 3.40 *c, f*).

Figure 3.40 shows that the light beam, similar to the superposition of two Gaussian beams, during propagation changes little of its appearance (changing only on a scale) and turns through an angle of $\pi/4$ (Fig. 3.40 *b, e*) and $5\pi/12$ (Fig. 3.40 *c, f*).

3.7.7. Trapping and rotation of microparticles in asymmetric Laguerre–Gauss beams

Further we describe the experiments, which prove that the OAM which the aLG beam transmits to a microparticle, and which is proportional to the velocity of the microparticles is also almost linearly dependent on the asymmetry parameter a. The scheme of the experiment on the manipulation of microparticles using asymmetric LG beams with linear polarization is shown in Fig. 3.41. The output beam of a solid-state laser L ($\lambda = 532$ nm, maximum output power 1500 mW) was expanded with the help of a system consisting of an MO_1 micro-objective (8", $NA = 0.3$) and an L_1 lens ($f_1 = 250$ mm). The expanded laser beam was directed to the display of a spatial light modulator SLM (PLUTO VIS, resolution 1920×1080 pixels, pixel size 8 µm). The modulated laser beam reflected from the modulator was sent through an L_2 ($f_2 = 350$ mm) and L_3 ($f_3 = 150$ mm) lens system to the MO_2 micro-lens entrance pupil (40", $NA = 0.65$). The image of the plane of manipulation was projected on the CMOS camera's LOMO TC-1000 matrix (resolution 1280×960 pixels, pixel size 1.67 µm) using an MO_3 micro-lens (16", $NA = 0.4$). To illuminate the trapping region, a system was used consisting of a condenser C and an illuminating lamp I, the light from which was entered by using the BS beam splitter. For manipulation in the experiment polystyrene microspheres with a diameter of 5 µm were used, which were located on a glass plate P, the radius of a Gaussian beam $w = 1$ mm.

Depending on the complex displacement parameter x_0 of the asymmetric LG beam formed in SLM, during the experiments either polystyrene microspheres rotated along the formed light ring

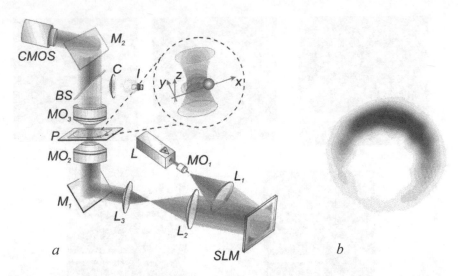

Fig. 3.41. Experimental setup for studying the transmission of OAM to polystyrene microspheres using asymmetric LG beams: L – solid-state laser ($\lambda = 532$ nm, power 1500 mW), MO_1 – micro-lens (8", $NA = 0.3$), L_1, L_2 and L_3 – lenses with focal lengths ($f_1 = 250$ mm, $f_2 = 350$ mm, $f_3 = 150$ mm), SLM – spatial light modulator PLUTO VIS (resolution 1920×1080 pixels, pixel size 8 µm), MO_2 – micro-lens ($40^\times$, $NA = 0.65$), P – glass plate with suspension of polystyrene microspheres, MO_3 – micro-lens (16", $NA = 0.4$), CMOS – video camera LOMO TC-1000 (resolution 1280×960 pixels, pixel size 1.67 µm), C – condenser, I – illuminating lamp, BS – beam splitter. An asymmetric LG beam (negative) (*b*) generated by SLM is shown with the following parameters: $w_0 = 1$ mm, $n = 8$, $x_0 = 0.2w_0$, $y_0 = 0.2w_0 i$, frame size 2×2 mm.

with an asymmetric intensity distribution (at $x_0 < 0.3w$), or trapped microspheres moved along the formed light curve with a crescent shape (at $x_0 \geq 0.3w$). When a particle moves along a long trajectory (for $x_0 < 0.3w$), it is influenced by a force proportional to the density of the OAM (3.165). And when a particle moves along a short trajectory (for $x_0 \geq 0.3w$), a force acts on it, which is proportional to the (full) OAM averaged along the trajectory.

Having appeared in the region of the aLG beam, the microsphere begins to experience the action of force that is directly proportional to the OAM of the beam. Under the action of this force, as well as the force arising from the intensity gradient, the microsphere begins to move uniformly accelerated along the formed light curve. The greater the density of the OAM, the greater the acceleration obtained by the microsphere when moving in the aLG beam. When an equilibrium of forces is set in motion, which causes the particle to move, and the forces of friction and resistance of the fluid acting

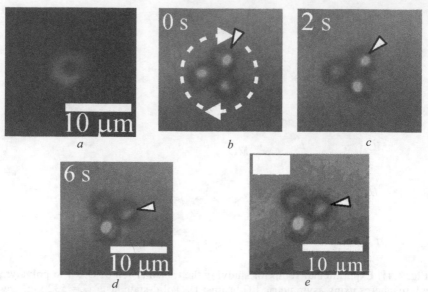

Fig. 3.42. The intensity distribution in the area of manipulation (leftmost frame) and the stage of movement of three polystyrene microspheres captured by an asymmetric Bessel beam with parameters $n = 3$, $x_0 = 0.1$ $w = -iy_0$.

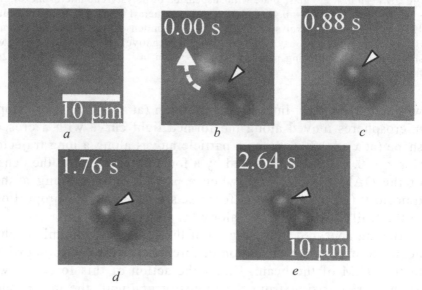

Fig. 3.43. The intensity distribution in the area of manipulation (the leftmost frame) and the stage of movement of the polystyrene microsphere trapped by an asymmetric Bessel beam with parameters $n = 3$, $x_0 = 0.4w = -iy_0$.

on the microsphere, it ceases to move at a uniform acceleration, and begins to move uniformly.

Figure 3.42 shows frames of rotation of three microparticles in the focus of an asymmetric LG beam, asymmetry parameters $x_0 = 0.1w = -iy_0$ ($n = 3$). The average linear velocity of rotation of the particles in the beam after the establishment of an equilibrium of all forces was equal to 0.45 ± 0.02 µm/s (Fig. 3.42).

Typical results of experiments using asymmetric LG beams with the parameter $x_0 = 0.4w = -iy_0$ are shown in Fig. 3.43. As can be seen, the stationary polystyrene microsphere, having appeared in the beam region, accelerates and moves along the light curve formed in the trapping plane (the average speed of movement is 0.84 ± 0.04 µm/s).

It should be noted that the aLG beams were formed using the SLM phase modulator, which reflected the same part of the energy for beams with different asymmetry parameter *a*. But only a fraction the total reflected energy reached the focus area, where microspheres are trapped. Moreover, the greater the asymmetry of the aLG beam, the less energy after the modulator falls into the focus region. Thus, the efficiency of the formation of the aLG beam presented in Fig.

Table 3.1. Calculated ratios of normalized OAMs of the asymmetric LG beams with parameters $x_0/w = 0, 0.1, 0.2, 0.3, 0.4$ and 0.5 to the OAMs of the asymmetric LG beam with the parameter $x_0 = 0.0$ ($n = 3$)

$x_0/w = -iy_0/w$	0.0	0.1	0.2	0.3	0.4	0.5
$\dfrac{J_z/W\vert_{x_0}}{J_z/W\vert_{x_0=0}}$	1.00	1.03	1.10	1.19	1.31	1.43

3.43, 1.8 times less than the efficiency of beam formation in fig. 3.42. And, nevertheless, the speed of movement of the microsphere along a shorter trajectory (Fig. 3.43) is almost 2 times higher than along a long trajectory (Fig. 3.42).

Table 3.1 shows the calculated values of the ratio of OAM of beams with different parameters $x_0/w = -iy_0/w$ to the OAM of the beam with the parameter $x_0 = 0$ ($n = 3$).

To experimentally confirm the existence of this dependence, experiments were carried out with asymmetric LG beams with the same parameters. We calculated the speed and acceleration of the microsphere, which it acquired some time *t* after the start of movement until the particle motion could still be considered as equally accelerated. In all experiments, time *t* was the same. Table 3.2 shows the increase in speed, and Table 3.3 – shows the acceleration

Table 3.2. Experimentally measured values of the ratios of the velocity of polystyrene microspheres trapped by asymmetric LG beams with the same topological charge $n = 3$ but with different values of the complex displacement parameter $x_0/w = -iy_0/w$

$x_0/w = -iy_0/w$	0.0	0.1	0.2	0.3	0.4	0.5
$\dfrac{v\big|_{x_0}}{v\big|_{x_0=0}}$	1.0	1.2±0.2	1.3±0.2	1.4±0.2	1.8±0.3	1.9±0.3

Table 3.3. Experimentally measured values of the ratios of acceleration values of polystyrene microspheres trapped by asymmetric LG beams with the same topological charge $n = 3$, but with different values of the complex displacement parameter $x_0/w = -iy_0/w$

$x_0/w = -iy_0/w$	0.0	0.1	0.2	0.3	0.4	0.5
$\dfrac{a\big|_{x_0}}{a\big|_{x_0=0}}$	1.0	1.2±0.3	1.3±0.3	1.3±0.3	1.4±0.3	1.6±0.3

of microspheres with increasing asymmetry parameter $a = x_0/w$. Note that the intensity in the light crescent is not constant (Fig. 3.41 *b*), therefore, the particle is affected by a gradient force, which leads first to acceleration when moving to the point with maximum intensity and to deceleration after its passage. Due to the symmetry of the intensity relative to the centre of the crescent, acceleration and deceleration compensate each other, so the gradient force was not taken into account when analyzing the experimental results.

Thus, it has been experimentally shown that the velocities of movement of microparticles trapped in asymmetric LG beams increase with increasing asymmetry parameter while maintaining the topological charge. The results of the calculation (Table 3.1) and the experiment (Tables 3.2 and 3.3) are consistent.

In this section we reviewed the generalization [187, 188] of the well-known Laguerre–Gauss laser modes. The asymmetric Laguerre-Gaussian beams do not have mode properties and have an asymmetric intensity distribution in a plane perpendicular to the direction of propagation. When an asymmetric LG beam propagates in a uniform space, the asymmetry of the first light ring decreases and the light energy is redistributed to the peripheral rings. The number of light rings coincides with the number of rings of the standard (symmetric) LG mode. The power of asymmetric LG beams and the projection of their OAM onto the optical axis are calculated analytically. It is established that the normalized OAM (OAM per photon) parabolically increases with an increase in the asymmetry parameter, which is equal to the ratio of the magnitude of the displacement along the Cartesian coordinate to the radius of the waist of the Gaussian beam. The conditions were obtained at which the normalized OAM coincides with the topological charge (as in the usual Laguerre–Gauss modes). As a special case, the aLG beams with zero radial index, having a cross-section of intensity in the form of a crescent, are considered. An expression for the coordinates of the intensity maximum is obtained and the rotation of the crescent during propagation in space is shown. The aLG beam with zero radial index was formed using a liquid-crystal spatial light modulator. Rotation of the intensity pattern in the beam cross section (crescent) upon propagation was experimentally demonstrated. The possibility of the formation of non-coaxial superpositions of the aLG beams with an intensity distribution close to Gaussian, rotating with propagation in space as a whole, is shown. Asymmetric crescent-shaped LG beams can be used for optical trapping and movement by biological objects (cells) [185], since the cell is less exposed to heat than when trapped in a symmetrical Gaussian beam. Also, the aLG beams will be useful in quantum communication systems for the formation of OAM-entangled photons, since the OAM of the aLG beam can be not only integral but also fractional, and the fractional OAM corresponds to the photon in the entangled state [186].

3.6. Asymmetrical Gaussian optical vortex

In [186] A. Zeilinger made it possible to form photons entangled in orbital angular momentum by displacing the centre of the waist of a Gaussian beam incident on a 'fork' hologram [97] that forms an optical vortex from the centre of this hologram. He suggested that in this case, a superposition of a Gaussian beam and a first-order Laguerre–Gauss beam propagates after the hologram. In such a superposition, the OAM will be fractional, that is, the photons will be in an entangled state.

When the centres of the incident Gaussian beam and the spiral phase plate (or hologram with a 'fork') are shifted, a beam is formed with an intensity distribution in the form of a crescent. Beams of the same type are formed if ordinary Bessel and Laguerre–Gauss beams make a complex-valued shift along both Cartesian coordinates. Such beams are called asymmetric Bessel beams [87, 88, 123, 124] and asymmetric Laguerre–Gauss beams [187, 188].

In this section, we obtain the analytical expressions for the complex amplitude of the laser beam transmitted by the SPP (spiral phase plate) if the beam incident on it is shifted from the optical axis. We also obtain the OAM of such an asymmetric Gaussian optical vortex (AGV).

3.8.1. The amplitude of the asymmetric Gaussian optical vortex

Consider a Gaussian beam with an amplitude in the waist in the form

$$E_0(x,y) = \exp\left(-\frac{x^2 + y^2}{w^2}\right), \tag{3.193}$$

where w is the waist radius. Let an optical vortex with a topological charge n, shifted by a distance x_0 along the x axis from the centre of the Gaussian beam, be introduced into the beam (3.193). The amplitude of such an optical vortex is described by the expression:

$$A_n(x,y) = \left[\frac{(x - x_0) + iy}{w}\right]^n. \tag{3.194}$$

Then, using the Fresnel transform, one can obtain the amplitude of such an asymmetric Gaussian vortex at a distance z from the initial plane (the plane of the waist of the Gaussian beam) in cylindrical coordinates:

$$E_n(r,\varphi,z) = w^{-n}\left[q(z)\right]^{-(n+1)}\left(re^{i\varphi} - q(z)x_0\right)^n \exp\left[-\frac{r^2}{w^2(z)} + \frac{ikr^2}{2R(z)}\right], \tag{3.195}$$

where $z_0 = kw^2/2$ is the Fresnel length , k is the wave number of light,

$$w^2(z) = w^2\left(1 + \frac{z^2}{z_0^2}\right), \quad R(z) = z\left(1 + \frac{z_0^2}{z^2}\right), \quad q(z) = 1 + \frac{iz}{z_0}. \tag{3.196}$$

From (3.195) it can be seen that the amplitude of the AGV is described by two factors, the last of the factors in (3.195) is a radially symmetric Gaussian beam at a distance z with the waist radius of the beam $w(z)$ and the curvature radius of the wave front $R(z)$. And the factor in (3.195), responsible for the asymmetry of the beam, we write out separately:

$$F = \left(re^{i\varphi} - q(z)x_0\right)^n = \left[(r\cos\varphi - x_0) + i\left(r\sin\varphi - \frac{zx_0}{z_0}\right)\right]^n. \qquad (3.197)$$

The distribution of the field intensity (3.195) will be proportional to the magnitude of the squared value (3.197) and will look like:

$$I = |E_n(r,\varphi,z)|^2 = w^{-2n}|q(z)|^{-2(n+1)} \times$$

$$\times \left[r^2 + |q(z)|^2 x_0^2 - 2rx_0\left(\cos\varphi + \frac{z}{z_0}\sin\varphi\right)\right]^n \exp\left[-\frac{2r^2}{w^2(z)}\right]. \qquad (3.198)$$

From (3.198) it is seen that for $z = 0$, the factor (3.197) has the form

$$|F|^2 = \left[r^2 + x_0^2 - 2rx_0\cos\varphi\right]^n. \qquad (3.199)$$

The expression (3.199) together with the Gaussian exponent from (3.198) describes an inhomogeneous crescent-shaped ring at the centre of which at the point $x = x_0$ there is an isolated n^{th} order intensity zero. The crescent is positioned so that the maximum intensity on it (at a fixed radius r) is reached at $\varphi = \pi$, and the minimum at $\varphi = 0$. At a distance from the waist, equal to the Fresnel length $z = z_0$, instead of (3.199), we obtain for the factor (3.197) responsible for the beam asymmetry:

$$|F|^2 = \left[r^2 + 2x_0^2 - 2\sqrt{2}rx_0\cos(\varphi - \pi/4)\right]^n. \qquad (3.200)$$

From (3.200) it follows that at a distance $z = z_0$ the crescent turned 45 degrees counterclockwise. Maximum intensity on the crescent is on a line at the angle $\varphi = 5\pi/4$, while the minimum is on a line at the angle $\varphi = \pi/4$. When moving a long distance $z \gg z_0$ from the rearrangement instead of (3.200) we get:

$$|F|^2 = \left[r^2 + \left(\frac{zx_0}{z_0}\right)^2 - 2\frac{zx_0}{z_0}r\sin\varphi\right]^n. \qquad (3.201)$$

From (3.201) it can be seen that in the far diffraction zone the crescent rotates 90 degrees counterclockwise, and the intensity maximum lies on the beam at $\varphi = -\pi/4$, and the minimum of intensity at $\varphi = \pi/4$.

From (3.197) it directly follows that the isolated zero intensity of the AGV has the coordinates:

$$\begin{cases} x = x_0, \\ y = x_0 \dfrac{z}{z_0}. \end{cases} \tag{3.202}$$

That is, an isolated zero lying in the initial plane on the horizontal axis at the point $x = x_0$ with increasing distance z moves perpendicularly of the horizontal axis of the on-axis to infinity.

3.8.2. *Fractional orbital angular momentum of AGV*

In [186], it was experimentally shown that the displacement of the centre of the Gaussian beam relative to the centre of the optical vortex led to the formation of photons entangled in the orbital angular momentum. That is, it was shown that the AGV should have a fractional OAM. We show that this is true. Since the OAM is preserved during the propagation of a laser beam, it can be calculated in any plane, for example, in the plane of the waist. It can be shown that the energy W of the AGV (3.195) and the projection of the OAM J_z on the z axis will be equal to:

$$W = \int_0^\infty \int_0^{2\pi} |E_n(r,\varphi,z=0)|^2 \, r dr d\varphi = \left(\frac{\pi w^2}{2} \right) \left(\frac{x_0}{w} \right)^{2n} \sum_{l=0}^n \left(C_n^l \right)^2 l! \left(\frac{w}{\sqrt{2}x_0} \right)^l, \tag{3.203}$$

$$J_z = -i \int_0^\infty \int_0^{2\pi} \overline{E}_n \frac{\partial E_n}{\partial \varphi} \, r dr d\varphi = n \left(\frac{\pi w^2}{2} \right) \left(\frac{x_0}{w} \right)^{2n} \sum_{l=0}^n C_n^{n-l} C_{n-1}^{n-l} l! \left(\frac{w}{\sqrt{2}x_0} \right)^{2l}, \tag{3.204}$$

where $C_n^l = n! / [l!(n-l)!]$ are binomial factors. From (3.203), (3.204) it follows that the normalized OAM of the asymmetric Gaussian vortex (3.195) will be equal to:

$$\frac{J_z}{W} = n \frac{\sum\limits_{l=0}^{n} C_n^{n-l} C_{n-1}^{n-l} l! \xi^{2l}}{\sum\limits_{l=0}^{n} \left(C_n^l\right)^2 l! \xi^{2l}}, \quad \xi = \frac{w}{\sqrt{2}x_0}. \tag{3.205}$$

From (3.205) it can be seen that the normalized OAM of the AGV is generally fractional. At $x_0 = 0$, it follows from (3.205) that the well-known result that the OAM of the Laguerre–Gauss mode [9] is equal to the topological charge n:

$$— \tag{3.206}$$

From (3.205) we find simpler expressions for $n = 1$ and $n = 2$, from which we can draw specific conclusions about the behaviour of the OAM of AGV. For example, when $n = 1$ from (3.205) we get:

$$\frac{J_{1z}}{W} = \frac{1}{1 + \left(\dfrac{\sqrt{2}x_0}{w}\right)^2}. \tag{3.207}$$

This expression coincides with that previously obtained in [110]. From (3.207) it follows that in the absence of a displacement $x_0 = 0$, the OAM is equal to 1, when the centre of the optical vortex is shifted from the centre of the Gaussian beam to the distance $x_0 = w/\sqrt{2}$ the OAM of the beam will be 1/2,

At $x_0 \to \infty$ the OAM tends to zero. Physically, this is understandable, since the optical vortex, with sufficient distance, no longer belongs to the Gaussian beam, but lies in the region of space where there is no light energy. When $n = 2$ from (3.205) we get:

$$\frac{J_{2z}}{W} = 2 \frac{\xi^2 + \xi^4}{1/2 + 2\xi^2 + \xi^4}. \tag{3.208}$$

From (3.208) it follows that in the absence of an offset, $x_0 = 0$ ($\xi \to \infty$) OAM is 2, with $x_0 = w/\sqrt{2}$ OAM is 8/7, and with a large displacement, $x_0 \to \infty(\xi = 0)$ OAM is zero.

Above, we considered asymmetric Gaussian optical vortices, which can be experimentally realized when a Gaussian beam is diffracted by an amplitude hologram with a characteristic fork at a singularity point [97]. Such a situation will be described by equations (3.193)–(3.195). However, the amplitude hologram with 'forks' has a low efficiency (a few percent). To increase the efficiency in the formation of AGV, another optical scheme will be used below, when a Gaussian beam falls on a spiral phase plate realized using a liquid crystal microdisplay. Such an experimental scheme can also be described analytically, although the formulas are more cumbersome.

3.8.3. Diffraction of a Gaussian beam on a shifted spiral phase plate

We examine the incidence of a Gaussian beam (3.193) onto a SPP with the number n, the centre of which is shifted from the centre of the Gaussian beam at the value x_0 and its transmission is described by the expression:

$$B_n(x,y) = \exp\left(in\,\mathrm{arctg}\left(\frac{y}{x-x_0} \right) \right) = \frac{[(x-x_0)+iy]^n}{\left[(x-x_0)^2 + y^2 \right]^{n/2}}. \tag{3.209}$$

At a distance z from the SPP, the amplitude of the light field of the AGV can be written using the Fresnel transform:

$$E_n(x',y',z) = \frac{-ik}{2\pi z}\exp\left(ik\frac{x'^2+y'^2}{2z} \right)\int\limits_{-\infty}^{\infty}\int\limits_{-\infty}^{\infty} E_0(x,y)B_n(x,y)\times$$

$$\exp\left(ik\frac{x'^2+y'^2}{2z} \right)\exp\left(-ik\frac{xx'+yy'}{z} \right)dxdy. \tag{3.210}$$

Integrals in (3.210) can be calculated using the reference integral [65]

$$\int\limits_0^{\infty} e^{-pr^2} J_n(cr)r\,dr = \frac{\sqrt{\pi t}}{2\sqrt{2}p}\exp(-t)\left[I_{(n-1)/2}(t) - I_{(n+1)/2}(t) \right], \tag{3.211}$$

where $t = c^2/(8p)$, f $J_n(x)$ and $I_n(x)$ are the Bessel function and the modified n-th order Bessel function.

Then instead of (3.210) we get:

$$E_n(x',y',z) = \frac{\sqrt{\pi}}{\sqrt{2}q(z)} \exp\left(-\left(\frac{x_0}{w}\right)^2 + ik\frac{x'^2+y'^2}{2z}\right)\left(\frac{iA-B}{\sqrt{A^2+B^2}}\right)^n \times$$

$$\times \sqrt{t}e^{-t}\left[I_{(n-1)/2}(t) - I_{(n+1)/2}(t)\right], \tag{3.212}$$

where

$$A = -\frac{kx'}{z} + \frac{2ix_0}{w^2}, \quad B = -\frac{ky'}{z}, \quad q(z) = 1 + \frac{iz}{z_0},$$

$$t = \frac{\mathrm{Re} + i\,\mathrm{Im}}{2|q(z)|^2}, \quad \mathrm{Re} = \frac{x'^2+y'^2}{w^2} - \left(\frac{x_0 z}{wz_0}\right)^2 - 2\left(\frac{x_0 x'}{w^2}\right), \tag{3.213}$$

$$\mathrm{Im} = \left(\frac{x^2+y^2}{w^2}\right)\left(\frac{z_0}{z}\right) - \left(\frac{x_0}{w}\right)^2\left(\frac{z}{z_0}\right) + \left(\frac{x_0 x}{w^2}\right)\left(\frac{z}{z_0}\right).$$

The factor with the nth-power in (3.212) is phase, and does not affect the intensity (its modulus is 1). The argument t in the Bessel function in (3.212) is complex, and therefore it is difficult to draw certain conclusions about the amplitude of the AGV. From (3.212) it is only clear that the argument t depends on the variable x' and on the magnitude of the displacement x_0. This means that the type of amplitude (3.212) will be asymmetrical, and the degree of asymmetry will be proportional to the offset x_0. It also shows that in the absence of an offset, $x_0 = 0$, the amplitude type coincides, up to a constant, with a radially symmetric Gaussian optical vortex [189]:

$$E_n(r,\varphi,z) = \frac{\sqrt{\pi}(-i)^n}{\sqrt{2}q(z)} \exp\left(in\varphi + \frac{ikr^2}{2z}\right) \tag{3.214}$$

$$\sqrt{t'}e^{-t'}\left[I_{(n-1)/2}(t') - I_{(n+1)/2}(t')\right],$$

where

$$t' = \frac{\left(\frac{r}{w}\right)^2\left(1+\frac{iz_0}{z}\right)}{2|q(z)|^2}.$$

Since in the experiment (the next section) measurement of the intensity of the AGV is taken in the focal plane of the lens, for comparison with the experiment, we obtain instead of (3.212) an expression for the amplitude of the AGV in the far-field diffraction $(z \gg z_0)$. From (3.213) we first obtain the expression for the argument t with $z \gg z_0$:

$$t' = \left(\frac{krw}{2\sqrt{2}f} \right)^2 - \left(\frac{x_0}{\sqrt{2}w} \right)^2, \tag{3.215}$$

where f is the focal distance of a spherical lens. It can be seen from (3.215) that the argument of the Bessel functions and the Gaussian exponent in (3.213) do not depend on the polar angle φ. Therefore, the asymmetry of the intensity distribution at the focus of a spherical lens is only responsible for factor (3.213), which is similar to factor F in (3.197) and is equal to

$$F' = \left[\frac{iA - B}{\sqrt{A^2 + B^2}} \right]^n = \left[\frac{\dfrac{ky}{f} - \dfrac{2x_0}{w^2} - i\dfrac{kx}{f}}{\sqrt{\left(\dfrac{kx}{f} - i\dfrac{2x_0}{w^2} \right)^2 + \left(\dfrac{ky}{f} \right)^2}} \right]^n. \tag{3.216}$$

From (3.216) it can be seen that the factor depends on the variables x and y not symmetrically and therefore the dependence on the polar angle should appear in the intensity. Indeed, the field intensity (3.213) will be proportional to the modulus of the square of the factor F'

$$I_F = |F'|^2 = \left[\frac{\left(\dfrac{4x_0^2}{w^4} + \dfrac{k^2 r^2}{f^2} \right) - \dfrac{4kx_0 r \sin\varphi}{fw^2}}{\sqrt{\left(\dfrac{k^2 r^2}{f^2} - \dfrac{4x_0^2}{w^4} \right)^2 + \left(\dfrac{4kx_0 r}{fw^2} \right)^2 \cos^2\varphi}} \right]^n. \tag{3.217}$$

From (3.217) it can be seen that the minimum of the numerator (with a constant r) is reached at $\varphi = \pi/2$, and the maximum at $\varphi = -\pi/2$. This means that the crescent-shaped diffraction pattern turned 90 degrees counterclockwise when the AGV propagated from the SPP plane to the focal plane (since in the waist the zero intensity was on the horizontal axis at $x = x_0$). In this case, the coordinates of an isolated intensity zero in the focal plane are equal to:

$$\begin{cases} x = 0, \\ y = x_0 \dfrac{f}{z_0}. \end{cases} \qquad (3.218)$$

The coordinates on the y-axis of the intensity zero in (3.202) and (3.218) coincide if we replace z with f. The expressions (3.201) and the numerator in (3.217) also coincide if we again replace z with f and multiply the numerator (3.217) with $(f/k)^2$. Thus, we have shown that, qualitatively, the behaviour of the AGV coincides with the diffraction of a Gaussian beam on a shifted SPP (3.209) and an amplitude hologram with a 'fork' (3.194).

3.8.4. Experimental formation of AGV

Figure 3.44 shows the experimental scheme. Radiation from a solid-state laser with a wavelength of 532 nm was directed through the pinhole PH and the lens L_1 to the display of a spatial light modulator SLM, with the phase transmission function of a given order. The reflected beam from the modulator subjected to spatial filtering using a lens system L_2 and L_3 and the diaphragm D. Next, the filtered laser beam was directed to the lens L_4, which focused it on

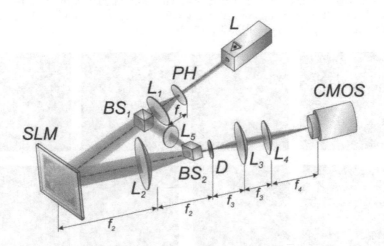

Fig. 3.44. Experimental scheme: L – solid-state laser (λ = 532 nm), PH – pinhole (40 µm), L_1, L_2, L_3, L_4 and L_5 – lenses with focal lengths (f_1 = 150 mm, f_2 = 350 mm, f_3 = 150 mm, f_4 = 250 mm, f_5 = 150 mm), BS_1, BS_2 – beam-splitting cubes, SLM – spatial light modulator PLUTO VIS (resolution 1920 × 1080 pixels, pixel size is 8 µm), D is the aperture that plays the role of a spatial filter, CMOS is a video camera ToupTek U 3 CMOS 08500 KPA (pixel size 1.67 µm).

the matrix CMOS camera. To obtain interferograms, beam-splitting cubes BS_1 and BS_2 were added to the scheme. The first of them divided the original beam into two, one of which was directed to the light modulator, and the second remained unchanged. Further, these two beams were combined to one using the second beam-splitting cube, so that a picture of their interference could be observed on the

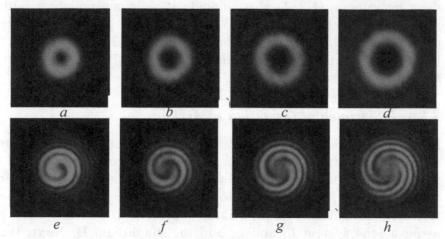

Fig. 3.45. The centre of the incident beam and the centre of SPP coincide ($x_0 = 0$). Top row – the obtained intensity distributions for SPP with a topological charge: (*a*) $n = 1$, (*b*) $n = 2$, (*c*) $n = 3$, (*d*) $n = 4$. The bottom row is the interferograms obtained for SPP with a topological charge: $n = 1$ (*e*), $n = 2$ (*f*) $n = 3$ (*g*), $n = 4$ (*h*). The sizes of pictures are 1200 by 1200 microns.

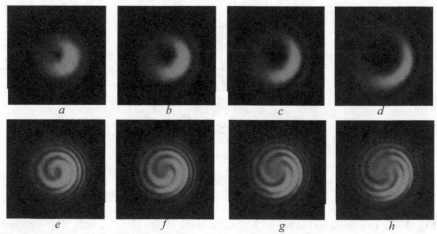

Fig. 3.46. The displacement of the centre of the incident beam from the centre of SPP is $x_0 = 0.250w$. Top row – the intensity distributions obtained for for SPP with topological charge: (*a*) $n = 1$, (*b*) $n = 2$, (*c*) $n = 3$, (*d*) $n = 4$. Lower row – the interferograms obtained for SPP with topological charge: $n = 1$ (*e*) $n = 2$ (*f*). $n = 3$ (*g*), $n = 4$ (*h*). The dimensions of the pictures are 1200 by 1200 μm.

camera. The L_5 lens was used to introduce a spherical wave front beam into the Gaussian beam. The waist diameter of the Gaussian beam is $\omega = 1400$ μm, the wavelength is $\lambda = 532$ nm.

The experiments investigated the effect of the displacement of the centre of the illuminating Gaussian beam relative to the centre of the spiral phase plate (SPP) displayed on the display of the light modulator. Figures 3.45 and 3.46 show the intensity distributions formed in these cases in the focal plane of the lens with a focal length of $f = 250$ mm. Also shown are the corresponding interferograms obtained as a result of the interference of the beams under study and the Gaussian beam with a spherical wave front.

Figure 3.45 shows the intensity distributions (*a–d*) and their corresponding interferograms (*e–h*) for optical Gaussian vortices with topological charges *n* from 1 to 4 with no displacement between the centres of the SPP and the Gaussian beam. It is seen from Fig.3.45 that the pictures of optical vortices have radial symmetry, that is, well-known doughnut-type pictures are formed with zero intensity in the centre.

Figure 3.46 shows the intensity distributions of Gaussian optical vortices with topological charges *n* from 1 to 4 with a large displacement of the centre of SPP from the centre of the Gaussian beam ($x_0 = 0.25w$). From Fig. 3.46 it is clear that 'crescents' formed instead of 'doughnuts'.

In this section, the asymmetrical Gaussian optical vortices are formed theoretically and experimentally that are formed during the diffraction of a Gaussian beam on a spiral phase plate or amplitude hologram with a 'fork' shifted from the optical axis. It is shown that, depending on the magnitude of the displacement, the laser beam has the form of a crescent, which rotates during propagation. An analytical expression is obtained for the orbital angular momentum of such a beam, which turns out to be fractional. Moreover, as the displacement increases, the momentum decreases the faster the higher the helicity number of the phase plate or 'fork' hologram. The experimental results are qualitatively consistent with theory and numerical simulation.

Focussing of vortex laser beams

4.1. Pearcey beams with autofocussing

The phenomenon of diffraction is one of the manifestations of the wave nature of light. When the light beam propagates along the optical axis from the initial plane to the observation plane (both planes are parallel to each other and perpendicular to the optical axis), the light fields coming from different parts of the initial plane interfere with each other, and a diffraction pattern appears in the observation plane the intensity distribution in which is generally different from the distribution in the initial plane. The propagation of a light beam in a homogeneous medium is described by the Helmholtz equation and its paraxial approximation – a Schrödinger type equation. Despite diffraction, these equations have solutions that describe light fields that are free from this phenomenon. First of all, these are the well-known traditional and recently discovered Bessel [1, 78, 88], Mathieu [4] and Airy [18] asymmetric beams. The Bessel beams propagate without diffraction in the three-dimensional space, and the Airy beams – in the two-dimensional space. Also, the plane waves do not have diffraction, since the intensity distribution of them does not change as it propagates from one plane to another. In the three-dimensional case, any light field does not have diffraction, the angular spectrum of plane waves of which is non-zero on an infinitely thin ring [1]. Along with the Bessel and Airy beams, paraxial structurally stable beams that are invariant to propagation are of interest. Such beams are not diffraction-free, but during propagation the structure of their intensity distribution in the transverse plane does not change, only the scale changes. The best-known such beams are the Hermite–Gauss and Laguerre–Gauss beams [2], hypergeometric modes [6]. In a recent paper [8], Pearcey beams were considered as three-dimensional analogues of the Airy beams. The distribution of the complex amplitudes of such beams is described by the Pearcey function [190, 191], defined as the integral of the complex exponent, whose argument is a polynomial (like the Airy function). The angular spectrum of such beams is a phase modulated parabola. These beams have the autofocus feature and are restored after distortion by obstacles. A recent paper [192] proposed a virtual source that forms the Pearcey beam.

In this section we generalize Pearcey functions and considere structurally stable half Pearcey beams (HP beams). Normal Pearcey beams [8] are the sum of two first-order HP beams. The angular spectrum of the HP beams is not parabolic (as in the Pearcey beams in [8]), but only one half of it. Two-dimensional analogues of the Pearcey beams with accelerated (curved) trajectory are also considered.

4.1.1. Pearcey three-dimensional half beams

The complex amplitude of the Pearcey paraxial light beams [8] in the initial plane is:

$$E(x,y,z=0)=\text{Pe}\left(\frac{x}{x_0},\frac{y}{y_0}\right)=\int_{-\infty}^{+\infty}\exp\left[is^4+is^2\left(\frac{y}{y_0}\right)+is\left(\frac{x}{x_0}\right)\right]ds. \quad (4.1)$$

It was shown in [8] that during propagation the beam structure does not change, only a shift of the beam centre along one Cartesian coordinate and scaling along both Cartesian coordinates occurs, and the scale along the x and y axes is different:

$$E(x,y,z)=\frac{1}{(1-z/z_e)^{1/4}}\text{Pe}\left(\frac{x}{x_0(1-z/z_e)^{1/4}},\frac{y-y_0z/(2kx_0^2)}{y_0(1-z/z_e)^{1/2}}\right), \quad (4.2)$$

where $z_e = 2ky_0^2$.

At $z = z_e$, the complex amplitude has the form:

$$E(x,y,z_e)=\sqrt{\frac{i\pi}{y/y_0-(y_0/x_0)^2}}\exp\left[\frac{-iy_0x^2}{4(x_0^2y_0-y_0^3)}\right]. \quad (4.3)$$

At the end of [8], a generalization of the beams (4.1) is presented in the form of beams with a complex amplitude in the initial plane, described by the function

$$U(x,y) = \int_{-\infty}^{+\infty} \exp\left[i\left(s^{2n} + ys^n + xs^m\right)\right]ds, \qquad (4.4)$$

moreover, when $n = 2m$, such beams are structurally stable. The numbers n and m must be integers. The requirement of integerness follows from the fact that the integral in (4.4) is taken along the whole numerical axis; therefore, for fractional values of n and m in the interval $s < 0$, the exponent can become real and tends to infinity as $s \to -\infty$.

The expression (4.2) from (4.1) is derived using the Fresnel transformation, the calculation of which uses the replacement of the integration variable, and the limits of integration do not change, since they are infinite. However, replacing one of the limits by zero, when changing variables, the limits also do not change, so along with (4.1) and (4.4) one can consider a half-order Pearcey beam of the v-th order, which also does not change its structure during propagation and which has a complex amplitude in the initial plane:

$$E(x,y,z=0) = \mathrm{HPe}_v\left(\frac{x}{x_0}, \frac{y}{y_0}\right) =$$
$$= \int_0^{+\infty} \exp\left[is^{4v} + is^{2v}(y/y_0) + is^v(x/x_0)\right]ds. \qquad (4.5)$$

In an arbitrary plane, the complex amplitude of such a beam will take the form:

$$E(x,y,z) = \frac{1}{(1-z/z_e)^{1/(4v)}} \mathrm{HPe}_v\left(\frac{x}{x_0(1-z/z_e)^{1/4}}, \frac{y - y_0 z/(2kx_0^2)}{y_0(1-z/z_e)^{1/2}}\right). \qquad (4.6)$$

Note that, in contrast to (4.4), in (4.5) the integer value of the parameter v is no longer required, since the integral is taken only for positive values of the integration variable s.

The diffraction pattern of the beams (4.6) in the transverse plane has the form of oblique light lines of the diffraction thickness (Fig. 4.1 *a*). Function $\mathrm{HPe}_v(x/x_0, y/y_0)$ was calculated in the same way as in [8], i.e. rotation of the integration contour in the complex plane was used, but instead of changing the variables $s \to s'\exp(i\pi/8)$, the replacement $s \to s'\exp(i\pi/(8v))$ was applied. In the calculation the following parameters were used: wavelength $\lambda = 532$ nm, $x_0 = y_0 = \lambda$, $v = 1$.

The angular spectrum of the HP beam of the v-th order is:

$$\tilde{E}(k_x, k_y) = \begin{cases} \dfrac{x_0 y_0}{p}(-k_x x_0)^{\frac{1-v}{v}} \exp\left(ik_x^4 x_0^4\right) \times \\ \quad \times \delta\left(k_x^2 x_0^2 + k_y y_0\right), k_x x_0 < 0, \\ 0, k_x x_0 \geq 0. \end{cases} \qquad (4.7)$$

Similar to [8], the angular spectrum of (4.7) is nonzero on the parabola $k_x^2 x_0^2 + k_y y_0$, but not on the whole parabola, but only on its half (for $k_x x_0 < 0$).

A Pearcey beam with an initial field (4.1) is a linear combination of first-order HP beams (4.5):

$$\mathrm{Pe}\left(\frac{x}{x_0}, \frac{y}{y_0}\right) = \mathrm{HPe}_1\left(\frac{x}{x_0}, \frac{y}{y_0}\right) + \mathrm{HPe}_1\left(-\frac{x}{x_0}, \frac{y}{y_0}\right). \qquad (4.8)$$

This is also evident from Fig. 4.1a, b: the Pearcey beam from [8] is the result of the interference of the HP beam from Fig. 4.1 a and its mirror reflection with respect to the plane $x = 0$. Note that the terms in (4.8) describe two beams, with the propagation of which their amplitudes acquire only a factor depending on $y_0{}^2$, i.e. same for both beams. Therefore, the coefficients in a linear combination can be arbitrary and, nevertheless, the beam will retain its structure. In particular, we can consider the difference in (4.8) instead of the sum, and then we get a beam similar to the Pearcey beam in [8], but there will be a minimum in the centre instead of the intensity maximum. The intensity of the difference of two symmetric with respect to the plane $x = 0$ HP beams are shown in Fig. 4.1 c.

Figure 4.2 shows the same diffraction patterns as in Fig. 4.1, but for the case of fractional order $v = 1.5$.

Figure 4.2 also shows that with increasing order of the Pearcey function v the diffraction pattern did not change fundamentally, although at $v = 1.5$ the intensity peaks fall off faster to the edge of the picture.

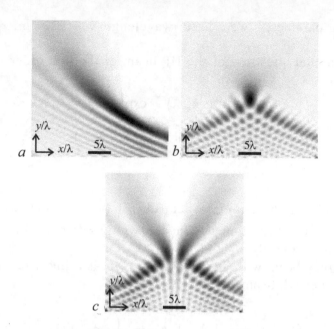

Fig. 4.1. The intensity distribution (negative) of the HP beam at $v = 1$ (a), the full Pearcey beam from [8] (b) and the difference of symmetric HP beams with zero intensity in the plane $x = 0$ (c).

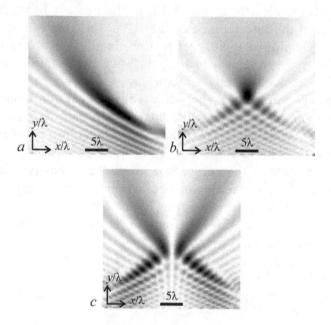

Fig. 4.2. The intensity distribution (negative) of the HP beam at $v = 1.5$ (a), the full Pearcey beam from [8] (b), and the difference between the half HP beams with zero intensity in the plane $x = 0$ (c).

Note that when obtaining (4.6) from (4.5), the integration variable can be changed only for $z < z_e$, otherwise the limits become imaginary. However, it can be shown that

$$E\left(x, \frac{y_0^3}{x_0^2} + \eta, z_e + \zeta\right) = E^*\left(-x, \frac{y_0^3}{x_0^2} - \eta, z_e - \zeta\right), \qquad (4.9)$$

thus, like a Pearcey beam [8] after passing a beam through the plane $z = z_e$ the intensity distribution is a mirror image of the intensity distribution to the plane $z = z_e$ with respect to the planes $x = 0$ and $y = y_0^3/x_0^2$.

In conclusion of this section, we note that the complex amplitudes of two three-dimensional HP beams with parameters (x_{01}, y_{01}) and (x_{02}, y_{02}) are orthogonal, provided that $y_{02}/y_{01} \neq x_{02}^2/x_{01}^2$. This condition can be obtained directly using the product of the functions $HPe_v(x/x_{01}, y/y_{01})$ and $HPe_v(x/x_{02}, y/y_{02})$, and from the fact that under this condition, the angular spectra of the plane waves (4.7) are different from zero on different non-intersecting parabolas.

4.1.2. Two-dimensional half Pearcey beams

Similarly to (4.5), we can consider 2D HP beams, whose complex amplitude in the initial plane has the form:

$$E(x,0) = \int\limits_{0}^{+\infty} \exp\left[is^p\left(\frac{x}{x_0}\right) + is^{2p}\right] ds = HPe_p^{2D}\left(\frac{x}{x_0}\right). \qquad (4.10)$$

Using the Fresnel transform, it can be shown that at some distance z the complex amplitude becomes

$$E(x,z) = \int\limits_{0}^{+\infty} \exp\left[is^p\left(\frac{x}{x_0}\right) + i\left(1 - \frac{z}{2kx_0^2}\right)s^{2p}\right] ds. \qquad (4.11)$$

Let us make the change of the integration variable $s = (1-z/z_e)^{-1/(2p)}t$, where $z_e = 2kx_0^2$. Then, instead of (4.11), we obtain for a complex amplitude of 2D HP beam:

$$E(x,z) = \frac{1}{(1 - z/z_e)^{1/(2p)}} HPe_p^{2D}\left(\frac{x}{x_0(1 - z/z_e)^{1/2}}\right). \qquad (4.12)$$

As in the three-dimensional case, here the replacement of the integration variable can be done only for $z < z_e$. Otherwise, the limits become imaginary. But from (4.11) it follows that

$$E(x, z_e + \zeta) = E^*(-x, z_e - \zeta), \qquad (4.13)$$

i.e. after passing the 2D HP beam through the plane $z = z_e$ its intensity distribution is a mirror image of the intensity distribution to the plane $z = z_e$ relative to the optical axis.

From (4.12) it can be seen that at the beginning of the propagation (to the focal plane $z = z_e$) in any transverse plane $z < z_e$, the coordinates of the intensity maxima of the 2D HP beam are determined by the formula:

$$x_{\max} = y_m x_0 \sqrt{1 - \frac{z}{z_e}}, \qquad (4.14)$$

where y_m is the coordinate of the m-th maximum of the function $\left| \text{HPe}_p^{2D}(x) \right|^2$. Twice differentiating (4.14) with respect to z, we obtain that for $z < z_e$, the inequality $(dx_{\max}/dz)(d^2 x_{\max}/dz^2) > 0$ will hold, which indicates the presence of an acceleration of the beam trajectory before focussing. From the property of symmetry (4.13), it follows that after passing through the $z = z_e$ plane, the 2D HP beam begins to propagate with deceleration.

Consider the propagation of the 2D HP beam when $p = 2$. Let $\lambda = 532$ nm, $x_0 = \lambda$. Then $z_e = 4\pi\lambda$. At a distance $z = 3z_e/4$ the beam is narrowed twice, in our case this distance is equal to $z = 3\pi\lambda$. The maximum intensity is doubled. In the plane $z_e = 4\pi\lambda$ the field (4.12) is 'infinitely' narrowed, and the focus arises. For the selected parameters, the intensity in the initial plane has the form shown in Fig. 4.3 *a*. The simulation was performed by the finite difference beam propagation method (BPM-method) in the region $-80\lambda \leq x \leq 80\lambda$, $0 \leq z \leq 30\lambda$. Figure 4.3 *b* shows the calculated intensity distribution in the plane Oxz. Figures 4.3 *c* and *d* show the intensity sections (Fig. 4.3*b*) at $z = 0$ (Fig. 4.3 *c*), and $z = 75\lambda/8$ ($\approx 3\pi\lambda$) (Fig. 4.3 *d*). Figure 4.3 shows that actually the beam was narrowed twice: at $z = 0$ the coordinate of the fourth zero is approximately -10λ, and at $z = 75\lambda/8$ it is about -5λ. The maximum intensity in Fig. 4.3 *d* is twice as high as the maximum intensity in Fig. 4.3 *c*. Besides, it is visible that both in Fig. 4.3 *c* and in Fig. 4.3 *d* the intensity of the second maximum is approximately half the intensity of the main maximum, which also demonstrates the

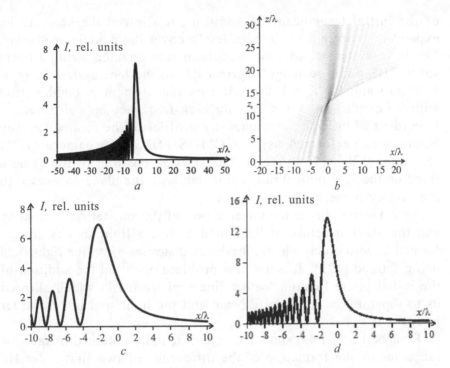

Fig. 4.3. The intensity distribution of the 2D HP beam, calculated by the BPM method: intensity cross section along the x axis at $z = 0$ (a), intensity in the Oxz plane (negative) (b), intensity cross section along the x axis at $z = 0$ (c) (enlarged picture (a)) and with $z = 75\lambda/8$ (d).

change in the light beam when it propagates only on a scale, but not in the structure.

In conclusion of this section, we note that, in contrast to the three-dimensional beams, the complex amplitudes of two 2D HP beams with the parameters x_{01} and x_{02} are not orthogonal.

4.1.3. Observation of the difference of two first-order HP beams

The Pearcey beams were produced using the optical scheme presented in Fig. 4.4. A spatial light modulator SLM PLUTO-VIS (resolution 1920 × 1080 pixels, pixel size 8 microns) was used. The output beam of a Laser solid-state laser ($\lambda = 532$ nm) was attenuated using filters of neutral density F. A system made from MO micro-lens (40×, $NA = 0.6$), lens L_1 ($f = 350$ mm) and pinhole PH (hole size 40 μm) was used to obtain a uniform Gaussian profile of the intensity of the initial laser beam. In addition, it allowed the beam to be expanded in order for it to completely cover the modulator display. The beam reflected from the modulator was directed, using a beam splitter BS and a rectangular prism RP, to the lens system L_2 ($f_3 =$ 350 mm) and L_3 ($f_2 = 150$ mm). This lens system in combination with the diaphragm D was used for high-frequency optical filtration. Recording of the transverse intensity profiles of the formed Pearcey beams was performed using a CMOS MDCE-5A camera (1/2", resolution 1280×1024 pixels). The starting position of the camera fixed on the Rail optical rails coincided with the plane connected to the display plane of the modulator.

As is known, due to the imperfection of the anti-reflective coating and the pixel structure of light modulators, a light spot is always formed in zero order, which introduces distortions into the light field being formed [193]. To solve this problem, we used the addition of the initial phase function with a linear phase mask, which allowed us to separate the formed HP beam and the light spot in zero order in space.

Figures 4.5 and 4.6 show the experimentally obtained images for the cases of the formation of the difference of two first-order HP beams at different distances from the modulator. The images in Fig. 4.5 were obtained in the case of using the coded phase distribution by the formula (4.8), which is shown in semitones in Fig. 4.5 *a*. And Fig. 4.6 shows the intensity distribution obtained in the case of using the uncoded phase distribution presented in Fig. 4.5 *a*. Note that during the phase output shown in Figs. 4.5 *a* and 4.6 *a*, on the modulator›s display, a linear phase was added to them, as mentioned above (in Figs. 4.5*a* and 4.6, but the linear phase is not shown).

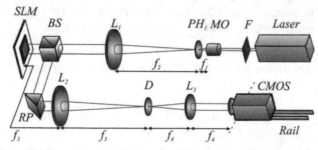

Fig. 4.4. Optical scheme for the formation of Pearcey light beams: Laser – solid-state laser ($\lambda = 532$ nm), F – filters of neutral density, MO – micro-lens (40×, NA = 0.6), PH – pinhole (40 microns), L_1, L_2 – lenses with focal length $f_2 = f_3 = 350$ mm, L_3 – lens with focal lengths $f_4 = 150$ mm, BS – beam divider, SLM – spatial light modulator PLUTO_VIS, RP – rectangular prism, D – diaphragm, CMOS – CMOS-camera MDCE-5A (1280x1024), Rail – optical rails.

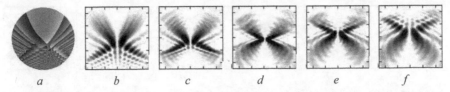

Fig. 4.5. The phase coded by formula (4.8) for forming the difference of two first-order HP beams (*a*) and experimentally formed intensity distributions (negatives) at different distances from the modulator: *b*) 50 mm, *c*) 250 mm; *d*) 400 mm; *e*) 550 mm; *f*) 650 mm. The grid spacing on images is 0.5 mm.

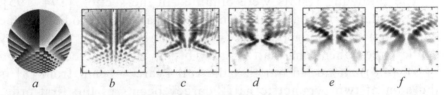

Fig. 4.6. Uncoded phase to form the difference of two first-order HP beams (4.5) (*a*) and experimentally formed intensity distributions (negatives) at different distances: *b*) 50 mm, *c*) 250 mm; *d*) 400 mm; *e*) 550 mm; *f*) 650 mm. The grid spacing on images is 0.5 mm.

From the presented images it is seen that the formed HP beams in both cases retain their structure during propagation in space. From the analysis of the images obtained, it follows that the formed beams have the autofocus property. The focus of the beams is at a distance of about 400 mm.

From the comparison of the figures it is clear that the intensity distributions in Fig. 4.5 are closer to the calculated distribution shown in Fig. 4.2 *c* than the intensity distributions in Fig. 4.6. It is also seen that in both cases the focus is at a distance of 400 mm, and after the focus, the intensity pattern is rotated 180 degrees with respect to the intensity pattern to the focus.

The experiments also demonstrated the self-restoration property of the formed HP beams. For this, a cover glass with an opaque area of 0.5 mm, applied using a black marker wa mounted on the beam path in the plane $z = 0$ (the initial position of the camera matrix) (Fig. 4.7 *a*). Thus, in this area, the absorption of light radiation occurred, which distorted the generated beam. Figure 4.7 presents the results of this experiment. From Fig. 4.7 *a* it is seen that the left upper maximum of the intensity of the formed beam is closed by an obstacle. In subsequent images, it is possible to observe how the restoration of the beam structure occurs; this is especially clearly seen in the images obtained after passing of the beam focus.

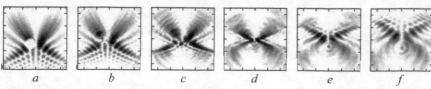

Fig. 4.7. Self-restoration of the generated difference of two first-order HP beams. Experimentally formed intensity distributions (negatives) at various distances: *a*) 0 mm, *b*) 100 mm; *c*) 250 mm; *d*) 400 mm; *e*) 550 mm; *f*) 700 mm. The grid spacing on the images is 0.5 mm.

The following results were obtained in the section [194, 195]. A new solution the paraxial Helmholtz equation (Schrödinger-type equations) describing the three-parameter family of structurally stable three-dimensional half Pearcey beams was obtained (equation (4.5), (4.6)). It turned out that the full Pearcey beam from [8] is the sum of two symmetric half Pearcey beams of the first order (Eq. (4.8)). With the help of a linear combination of two symmetric half Pearcey beams with different amplitudes it is possible to form new structurally stable beams that are neither complete Pearcey beams [8] nor half Pearcey beams. The formula for the amplitude of the angular spectrum of plane waves of the three-dimensional Pearcey half-beams was obtained (equation (4.7)); the amplitude of the spectrum is different from zero on the half of the parabola. The condition of orthogonality of the complex amplitudes of the three-dimensional half Pearcey beams is obtained: the half Pearcey beams are orthogonal and their Fourier spectrum of which lies on non-intersecting parabolas; it was established that the Pearcey two-dimensional beams are not orthogonal. The solution of the two-dimensional paraxial Helmholtz equation describing a two-dimensional analogue of the structurally stable half Pearcey beams (equations (4.10), (4.12)) was obtained. The two-dimensional half Pearcey beams, as well as the three-dimensional ones, have the property of autofocussing, as well as an accelerating trajectory to the focal plane (Fig. 4.3 *b*).

4.2. Hypergeometric autofocus beams

Vortex laser beams with orbital angular momentum are the subject of close attention in optics. They are used to increase the information capacity of the transmission channels in wireless communication systems with a speed of several terabits per second [196], for optical capture and rotation of microparticles [197], for atmospheric sensing in the presence of turbulence [198]. Examples of vortex beams are the well-known Bessel [1], Laguerre–Gauss [2] laser beams, and hypergeometric modes [6]. When propagating in space, these beams either retain their structure [1] or increase in scale [2, 6]. For use in micromanipulation tasks they need to be focused with the help of microlens. Such a focussing inevitably leads to a distortion of the beam shape due to aberrations of the focussing lens. However, laser beams with an autofocus feature are known. In such beams, the intensity distribution in the focus is strictly defined. Examples of laser beams with autofocussing are Pearcey beams [8]. Pearcey beams were considered in [8] as three-dimensional analogues of the Airy beams [18]. The distribution of the complex amplitudes of such beams is described by the Pearcey function [190, 191], defined as the integral of the complex exponent, whose argument is a polynomial (like the Airy function). The angular spectrum of such beams is a phase modulated parabola. These beams have the autofocus feature and are restored after distortion by obstacles. In [192] proposed a virtual source that forms the Pearcey beam. In [195, 199] a generalization of the Pearcey function is given and the structurally stable Pearcey half-beams are considered.

The ordinary Pearcey beams [8] are the sum of two first-order Pearcey beams. The angular spectrum of the half Pearcey beams is not one of the parabolas (as in the Pearcey beams from [8]), but only one half of it. Note that the Pearcey beams do not have an orbital angular momentum. Also in [195, 199], two-dimensional analogues of the Pearcey beams with a curved trajectory, that is, with acceleration during propagation, are considered.

In this section, on the basis of two-dimensional half-beams of Pearcey, the paraxial Pearcey vortex beams with orbital angular momentum are obtained. The complex amplitude of such beams is proportional to the degenerate hypergeometric function, and they coincide in shape with the hypergeometric modes [6]. But since the Pearcey beams have the autofocussing property, the Pearcey vortex beams obtained here also have the autofocus property.

4.2.1. Pearcey vortex beams or hypergeometric autofocussing beams

Consider the complex amplitude of a paraxial three-dimensional monochromatic laser beam, which is a continuous superposition of two-dimensional half Pearcey beams, each of which is rotated by the polar angle θ, and for each beam entering the superposition the phase is proportional to the angle of rotation $n\theta$:

$$E_{np}(x,y,z=0) = \frac{(-i)^n}{2\pi} \int_0^{2\pi} e^{in\theta} d\theta \times$$

$$\times \int_0^{\infty} \exp\left[is^{2p} + is^p \left(\frac{x\cos\theta + y\sin\theta}{x_0} \right) \right] ds, \qquad (4.15)$$

where n is an integer, which is called the topological charge of the optical vortex, p is a real number that determines the order of the half Pearcey beam. Note that in the ordinary Pearcey beams [8] $p = 2$, and the integral in (4.15) with respect to the variable s is taken from minus to plus infinity. The beams (4.15) with different topological charge n are orthogonal, and with different order p are not orthogonal.

Both integrals in (4.15) can be calculated, then instead of (4.15) we get:

$$E_{np}(r,\phi,z=0) = \frac{\exp\left[\dfrac{i\pi}{2}\left(\dfrac{pn+1}{2p} \right) + in\phi \right]}{2pn!} \times$$

$$\times \Gamma\left(\frac{pn+1}{2p} \right) \left(\frac{r}{2x_0} \right)^n {}_1F_1\left(\frac{pn+1}{2p}, n+1, -i\frac{r^2}{4x_0^2} \right), \qquad (4.16)$$

where (r, φ) are the polar coordinates, $\Gamma(x)$ is the gamma function, ${}_1F_1(a,b,x)$ is a degenerate hypergeometric function [65].

Using the Fresnel transform, one can obtain the complex amplitude of the Pearcey optical vortex (4.15) at an arbitrary distance z in the form:

$$E_{np}(r,\phi,z) = \frac{\exp\left[\dfrac{i\pi}{2}\left(\dfrac{pn+1}{2p} \right) + in\phi \right]}{2pn!\left(1 - \dfrac{z}{z_e} \right)^{1/2p}} $$

$$\times \Gamma\left(\frac{pn+1}{2p} \right) \xi^n \, {}_1F_1\left(\frac{pn+1}{2p}, n+1, -i\xi^2 \right), \qquad (4.17)$$

where

$$\xi = \frac{r}{2x_0\sqrt{1-\dfrac{z}{z_e}}}, \quad z_e = 2kx_0^2.$$

It can be verified that from (4.17) with $z = 0$ we get (4.16). From (4.17) it follows that for $n \neq 0$ on the optical axis ($r = 0$) for any z there will be zero intensity, and the intensity distribution will have the form of a set of concentric rings centered on the optical axis. But at $n = 0$ the beam (4.17) will not be vortex and circular, that is, the intensity on the optical axis will be non-zero, and the complex amplitude will be:

$$E_{0p}(r,z) = \frac{\exp\left(\dfrac{i\pi}{4p}\right)}{2p\left(1-\dfrac{z}{z_e}\right)^{1/2p}} \Gamma\left(\frac{1}{2p}\right) {}_1F_1\left(\frac{1}{2p}, 1, -i\xi^2\right). \tag{4.18}$$

The complex amplitude (4.18) describes a paraxial wave with infinite energy, converging to a point on the optical axis, similarly to spherical or parabolic waves. From (4.17) it can be seen that for $z = z_e$ the argument ξ tends to infinity. Using asymptotics as $x \to \infty$

$$x^n \left| {}_1F_1\left(\frac{pn+1}{2p}, n+1, -ix^2\right) \right| \approx \frac{1}{x}, \tag{4.19}$$

we find that in the autofocusing plane ($z = z_e$) the complex amplitude (4.17) is proportional to

$$E_{np}(r,\phi,z=z_e) \approx \frac{\exp(in\phi)}{r}. \tag{4.20}$$

From (4.20) it follows that in the autofocussing plane ($z = z_e$) a sharp focus is formed, but at the centre of the focus at one point only the intensity is zero (if $n \neq 0$). Let us briefly prove the presence of an axial zero intensity in the focal plane. The complex amplitude of the two-dimensional Pearcey beam in the focal plane is equal to

$$E(x,z=z_e) = \int\limits_{0}^{+\infty} \exp\left[is^p\left(\frac{x}{x_0}\right)\right] ds, \tag{4.21}$$

and the superposition (4.15) in this plane has the form (in polar coordinates):

$$E_{np}\left(r,\varphi,z=z_e\right)=\frac{(-i)^n}{2\pi}\int_0^{2\pi}e^{in\theta}d\theta\int_0^{\infty}\exp\left[is^p\,\frac{r\cos(\theta-\varphi)}{x_0}\right]ds, \qquad (4.22)$$

The internal integral on the optical axis ($r = 0$) is equal to infinity, but when changing the order of integration, we get:

$$E_{np}\left(r,\varphi,z=z_e\right)=\frac{(-i)^n}{2\pi}\int_0^{\infty}ds\int_0^{2\pi}\exp\left[in\theta+i\frac{rs^p}{x_0}\cos(\theta-\varphi)\right]d\theta=$$

$$=\exp(in\varphi)\int_0^{\infty}J_n\left(\frac{rs^p}{x_0}\right)ds. \qquad (4.23)$$

When $r = 0$, the integrand is strictly zero, and hence the axial intensity in the focal plane is also zero.

A remarkable property of the Pearcey vortex beams is that the autofocussing distance $z_e = 2kx_0^2$ does not depend on the parameters of the beam p and n. This means that any superposition of such beams with different numbers p and n will be focused at the same distance z_e. This property of vortex beams can be used to form a given intensity distribution and orbital angular momentum in the autofocussing plane.

Figure 4.8 shows the intensity distributions of the radial ($n = 0$) and vortex ($n = 5$) Pearcey beams at different distances from the initial plane: $z = 0$ (a, d), $z = 0.75z_e$ (b, e), $z = 0.999\,z_e$ (c, f). At a distance $z = 3z_e/4$, the diffraction pattern is compressed twice as compared with the picture in the plane $z = 0$, and at a distance $z = 0.999z_e$ – $10\sqrt{3}$ times. The intensity cross section shown in Fig. 4.8b, shown in Fig. 4.9, the width of the light spot in half intensity decay in Fig. 4.9 amounted to 0.21λ.

Figure 4.8 was calculated by a formula similar to (4.15):

$$E_{np}\left(x,y,z\right)=\frac{(-i)^n}{2\pi}\frac{1}{\left(1-z/z_e\right)^{1/(2p)}}=$$

$$=\int_0^{2\pi}e^{in\theta}\,\mathrm{HPe}_2^{2D}\left(\frac{x\cos\theta+y\sin\theta}{x_0\sqrt{1-z/z_e}}\right)d\theta, \qquad (4.24)$$

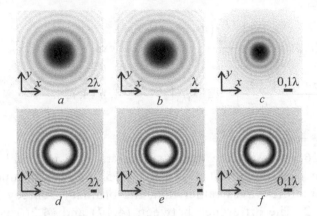

Fig. 4.8. Intensity distributions (negative) of the Pearcey vortex beam with $n = 0$ (*a, b, c*) and with $n = 5$ (*d, e, f*) in the planes $z = 0$ (*a, d*), $z = 3z_e/4$ (*b, e*), $z = 0.999z_e$ (*c, f*).

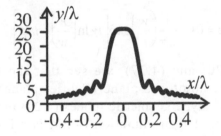

Fig. 4.9. The intensity cross section near the focus ($z = 0.999z_e$) with $n = 0$.

the identity connecting the Pearcey function with $p = 2$ with the Bessel function of the first kind was used:

$$
\text{HPe}_2^{2D}(\xi) = \frac{\pi}{4}\sqrt{\frac{|\xi|}{2}}\exp\left[-i\left(\frac{\xi^2}{8}-\frac{\pi}{8}\right)\right]J_{-1/4}\left(\frac{\xi^2}{8}\right) +
$$

$$
+i\frac{\pi}{4}\sqrt{\frac{|\xi|}{2}}\,\text{sgn}(\xi)\exp\left[-i\left(\frac{\xi^2}{8}-\frac{3\pi}{8}\right)\right]J_{1/4}\left(\frac{\xi^2}{8}\right).
$$

(4.25)

This identity is obtained from several reference and non-integrals (expressions 3.696 in [112]). We note that the Pearcey vortex beams (4.17) with $p = 1$ coincide in structure with the known hypergeometric modes [6] with $\gamma = 0$:

$$E_{n\gamma}(r,\phi,z) = \frac{\exp\left[-\dfrac{i\pi}{4}(n - i\gamma + 1) + in\phi\right]}{2\pi n!} \exp\left[\frac{i\gamma}{2}\ln\left(\frac{z}{z_0}\right)\right] \times$$

$$\Gamma\left(\frac{n + i\gamma + 1}{2}\right)\left(\frac{z_0}{z}\right)^{1/2} \times \xi^{n/2}\, {}_1F_1\left(\frac{n - i\gamma + 1}{2}, n + 1, i\xi\right),$$

(4.26)

wherein $\xi = kr^2/(2z)$, $z_0 = kw^2/2$.

In (4.26), w is a scale factor similar to x_0, γ is a real parameter similar to p. Although at $p = 1$ in (4.15) there is a superposition of the parabolic waves rather than the ordinary Pearcey beams [8] with $p = 2$. The difference between (4.17) and (4.26) is also that the Pearcey vortex beams (4.17) in the initial plane have the same form (4.16), and the hypergeometric modes (4.26) are generated by a complex amplitude of the form [6]:

$$E_{n\gamma}(r,\phi,z = 0) = \frac{1}{2\pi}\left(\frac{w}{r}\right)\exp\left[i\gamma\ln\left(\frac{r}{w}\right) + in\phi\right].$$

(4.27)

Comparing (4.20) and (4.27), we see that the Pearcey vortex beams (4.17) in the autofocus plane (Fourier plane) have the form (4.20), similar to the hypergeometric modes (4.27) in the initial plane (for $\gamma = 0$). We note that the function (4.20) is a Fourier invariant, and its Fourier transform is

$$E_n(\rho,\theta) = (-i)^{n+1}\frac{\exp(in\theta)}{\rho}, \rho > 0.$$

(4.28)

But with $n \neq 0$ and $\rho = 0$ the amplitude (4.28) is zero. It can be assumed that the hypergeometric modes (4.26) and the Pearcey vortex beams (4.17) differ in the optical axis shift by the autofocussing distance. This is indirectly confirmed by the methodology for obtaining accelerated beams from slowing down by displacement along the optical axis [66].

Equating the argument ξ in (4.17) to the value at which the amplitude modulus reaches a maximum ξ_0, we obtain the dependence of the change in the radius of the ring with maximum intensity on the distance along the optical axis:

$$r_0 = 2x_0\xi_0\sqrt{1 - \frac{z}{z_e}}.$$

(4.29)

From (4.29) it can be seen that at $z = z_e$ the radius of the ring is zero. If we calculate the first and second derivatives with respect to z from the radius of the ring (4.29), then they will be of the same sign. This means that the Pearcey vortex beam propagates to the focus along an accelerating trajectory and converges 'sharply' into focus, similarly to the Airy ring beams [64].

4.2.2. Pearcey–Gauss vortex beams

The beams (4.17), like the beams (4.26), have infinite energy, similarly to the Bessel beams [1]. It is clear that if the Fresnel transformation from function (4.15) is explicitly calculated using special functions (4.17), then the Fresnel transformation from function (4.15) multiplied by the Gaussian exponent can be calculated in the same way. Therefore, we consider the Pearcey–Gauss vortex beams in the initial plane, multiplying the complex amplitude (4.15) by the Gaussian exponent:

$$F_{np}(x,y,z=0) = \frac{(-i)^n}{2\pi}\exp\left(-\frac{x^2+y^2}{\sigma^2}\right)\int_0^{2\pi} e^{in\theta}d\theta \times$$
$$\times \int_0^\infty \exp\left[is^{2p} + is^p\left(\frac{x\cos\theta + y\sin\theta}{x_0}\right)\right]ds,$$

(4.30)

where σ is the radius of the waist of the Gaussian beam.

The complex amplitude of the Pearcey–Gauss vortex beams at an arbitrary distance from the initial plane is obtained using the Fresnel transformation and is equal to:

$$F_{np}(r,\varphi,z) = \frac{\exp\left[-\frac{i\pi}{2}(n+1)+in\varphi\right]}{2pn!}\exp\left[-\frac{r^2}{\sigma^2(z)}+\frac{ikr^2}{2R(z)}\right]\times$$
$$g^{(1-2p)/p}(z)q^{-1/p}(z)\times\Gamma\left(\frac{pn+1}{2p}\right)\xi^n\,{}_1F_1\left(\frac{pn+1}{2p},n+1,-i\xi^2\right).$$

(4.31)

where

$$\xi = \frac{r}{2x_0 g(z)q(z)}, \quad g(z) = \sqrt{\frac{z}{z_0}-i},$$

$$q(z) = \sqrt{\frac{z}{z_e}-i\left(\frac{z}{z_0}-i\right)}, \quad z_0 = \frac{k\sigma^2}{2},$$

$$\sigma(z) = \sigma\sqrt{1+\frac{z^2}{z_0^2}}, \; R(z) = z\left(1+\frac{z_0^2}{z^2}\right).$$

At $\sigma \to \infty$, the amplitude (4.31) goes into (4.17). Note that the product $g(z)q(z)$ in the denominator of the argument ξ is a complex number and is not equal to zero for any z. For $z = z_e$, the modulus of the product $g(z)q(z)$ reaches a minimum. That is, unlike the Pearcey vortex beams (4.17), the Pearcey--Gauss vortex beams (4.31) in the autofocussing plane the beam radius is minimal, but non-zero. But at the same time, the Pearcey–Gauss vortex beams have finite energy and can be formed using a liquid-crystal light modulator.

We also note that the beams (4.31) with $p = 1$ and at $z_e \to \infty$ coincide in their structure with hypergeometric beams [63] with $m = -1$, $\gamma = 0$. But the beams in [63] do not have the autofocussing property.

The following the results were obtained in this section [200, 201]. A new solution of the paraxial Helmholtz equation (Schrödinger-type equations) describing the two-parameter family of structurally stable three-dimensional Pearcey vortex beams (equation (4.17)). It turned out that the complex amplitude of the Piercey vortex beam coincides in structure with the complex amplitude of the hypergeometric modes [6], but, unlike them, it has the autofocussing feature. The Pearcey vortex beam propagates to the focus along an accelerating trajectory and converges 'sharply' into a focus, analogous to that of the Airy ring beams [66].

An explicit expression was also obtained for the complex amplitude of the Pearcey–Gaussian vortex beams (equation (4.31)) with finite energy, which have the autofocussing property, but the beam radius in focus is finite, in contrast to the Piercey vortex beams with infinite energy, whose beam radius in focus is zero.

4.3. Hypergeometric beams in a parabolic waveguide

In 2007, the paraxial hypergeometric modes (HyG modes) were considered [6]. A little later, hypergeometric–Gaussian laser beams were obtained on the basis of HyG modes [202]. In [63], a more general family of HyG beams was considered, which include both special cases of the HyG mode [6] and hypergeometric–Gaussian beams [202]. Laser HyG beams were experimentally formed using diffraction optical elements [203] and holograms synthesized on a computer [204]. Recently, analytical expressions describing the propagation of HyG-beams in a gradient hyperbolic medium have been obtained [205] and uniaxial crystal [206].

In this section, we obtain an expression describing the propagation of HyG-beams in a 3D gradient parabolic waveguide. It was shown that the light field of the HyG-beams is periodically repeated, and after each half period the Fourier spectrum is formed.

A new family of the mode solutions for the Helmholtz equation in a parabolic medium was also obtained. These solutions are proportional to Kummer functions and diverge at infinity. However, for some values of the parameters of the Kummer function, these solutions become finite and pass into the known Laguerre–Gauss modes (LG-modes).

Recently, interest in microoptics gradient elements has increased. In [207–209] the Luneberg gradient microlens was invsetigated. In [207, 208], experiments were carried out with a Luneberg planar lens, and in [209] a simulation of a Luneberg planar photonic crystal lens showed that the focus formed by it has a width of half-decay of intensity FWHM (full width at half maximum) = 0.44λ for $n_0 = 1.41$. In [210, 211] a planar lens was considered, and in [212] a three-dimensional Mikaelian lens with a gradient index of refraction in the form of a hyperbolic secant was studied. In [212], it is shown that the Mikaelian 3D gradient lens forms a focal spot with a diameter of half intensity FWHM = 0.40λ. In [213], a Mikaelian planar subwave binary lens was considered, which forms a focus with a width FWHM = 159 nm = $0.102\lambda = 0.35\lambda/n$, $n = 3.47$ – the refractive index of the lens.

In this section, the propagation of a Gaussian beam in a gradient parabolic fiber is considered. A section of this fiber of a certain length can be considered as a gradient parabolic (GP) lens. Simulation showed that the focal spot in such a lens has a diameter FWHM = 0.42λ, with $n_0 = 1.5$ it has a refractive index on the axis of the lens. An analytical expression has also been obtained for the radii of the jumps of the refractive index for a binary lens that approximates a 3D GP lens. The smaller diameter of the elliptical focal spot of such a binary parabolic lens calculated by the FDTD method is FWHM = 0.45λ.

4.3.1. Paraxial hypergeometric beams in a parabolic gradient medium

Consider a parabolic gradient medium with a refractive index of the form:

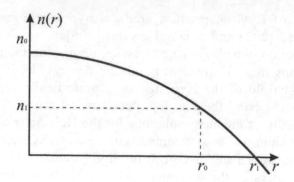

Fig. 4.10. Graph of the parabolic refractive index on the radial coordinate.

$$n^2(r) = n_0^2 \left[1 - \frac{(n_0^2 - n_1^2)}{n_0^2} \left(\frac{r}{r_0} \right)^2 \right], \tag{4.32}$$

where r is the radial transverse coordinate, n_0 and n_1 are the refractive indices on the optical axis ($r = 0$) and at $r = r_0$. Figure 4.10 shows the refractive index profile (4.32).

The paraxial equation of light propagation in a parabolic medium (4.32) has the form:

$$\left[2ik\frac{\partial}{\partial z} + \frac{\partial^2}{\partial x^2} + \frac{\partial^2}{\partial y^2} - (\alpha k)^2 \frac{(x^2 + y^2)}{4} \right] E(x,y,z) = 0, \tag{4.33}$$

where $\alpha = \dfrac{2\sqrt{n_0^2 - n_1^2}}{r_0 n_0}$, $k = \dfrac{2\pi}{\lambda} n_0$, λ is the wavelength. In [214], it was shown that the general solution (4.33) has the form:

$$E(x,y,z) = \frac{\alpha k}{2\pi i \sin(\tau z)} \exp\left[\frac{i\alpha k}{4\tan(\tau z)} (x^2 + y^2) \right] \times$$

$$\times \int\int\limits_{-\infty}^{\infty} E_0(\xi, \eta) \exp\left\{ \frac{i\alpha k}{4\sin(\tau z)} \left[(\xi^2 + \eta^2)\cos(\tau z) - 2(x\xi + y\eta) \right] \right\} d\xi d\eta, \tag{4.34}$$

where $\tau = \alpha/2$. Note that the integral transform (4.34), up to notation, coincides with the partial Fourier transform [215, 216]. In the cylindrical coordinates, the equation (4.34) for the initial field $E_0(r, \varphi) = E_0(r)\exp(in\varphi)$ is:

$$E(\rho,\theta,z) = (-i)^{n+1} \frac{2k}{f_2} \exp\left(\frac{ik\rho^2}{2f_1} + in\theta\right) \times$$

$$\times \int_0^\infty E_0(r) \exp\left(\frac{ikr^2}{2f_1}\right) J_n\left(\frac{kr\rho}{f_2}\right) r\,dr,$$

(4.35)

where n is an integer, $J_n(x)$ is the Bessel function,

$$f_1 = f_0 \tan(z / f_0),$$
$$f_2 = f_0 \sin(z / f_0),$$
$$f_0 = \frac{2}{\alpha} = \frac{r_0 n_0}{\sqrt{n_0^2 - n_1^2}}.$$

(4.36)

In [217], expressions describing transformations of LG beams in a parabolic fiber were obtained. Next, we consider the propagation of HyG beams in a parabolic fiber (4.32), which in the initial plane are described by a complex amplitude [63]:

$$E_0(r) = \left(\frac{r}{\delta}\right)^{m+i\gamma} \exp\left(-\frac{r^2}{2\sigma^2}\right),$$

(4.37)

where m, γ are real numbers, δ and σ are the scale of the power amplitude component and the radius of the Gaussian beam. Substitute expression (4.37) into (4.35) and use the reference integral [65]:

$$\int_0^\infty y^{\mu-1} \exp\left(-\beta y^2\right) J_n(cy)\,dy = \frac{c^n \beta^{-(n+\mu)/2} \Gamma\left(\frac{n+\mu}{2}\right)}{2^{n+1} n!} \times$$

$$\times {}_1F_1\left(\frac{n+\mu}{2}, n+1, -\frac{c^2}{4\beta}\right),$$

(4.38)

where $\Gamma(x)$ is a gamma function, ${}_1F_1(a,b,x)$ is the Kummer function [65], then we get instead of (4.35):

$$E_{m\gamma n}(\rho,\theta,z)=(-i)^{n+1}\left(\frac{2k\sigma^2}{f_2}\right)\left(\frac{\sqrt{2}\sigma}{\delta}\right)^{m+i\gamma}\times$$

$$\exp\left(\frac{ik\rho^2}{2f_1}+in\theta\right)\times$$

$$\times\left(1-\frac{ik\sigma^2}{f_1}\right)^{-(m+i\gamma+2)/2}\frac{\Gamma\left(\dfrac{n+m+i\gamma+2}{2}\right)x^{n/2}}{n!}\times \qquad (4.39)$$

$$\times {}_1F_1\left(\frac{n+m+i\gamma+2}{2},n+1,-x\right),$$

where

$$x=\frac{\rho^2}{2\omega^2(z)}+\frac{ik\rho^2}{2R(z)},$$

$$\omega(z)=\sigma\left[\cos^2\left(\frac{z}{f_0}\right)+\left(\frac{f_0}{k\sigma^2}\right)^2\sin^2\left(\frac{z}{f_0}\right)\right]^{1/2}, \qquad (4.40)$$

$$R(z)=\frac{1}{2}f_0\sin\left(\frac{2z}{f_0}\right)\left[1+\left(\frac{f_0}{k\sigma^2}\right)^2\tan^2\left(\frac{z}{f_0}\right)\right].$$

The modular light field (4.39) will repeat with a period $L = \pi f_0$. The amplitudes in the planes, separated by a distance of half a period $L_1 = (\pi f_0)/2$, are related by the Fourier transform:

$$E_{m\gamma n}(\rho,\theta,z=L_1)=(-i)^{n+1}\left(\frac{k\sigma^2}{f_2}\right)\left(\frac{\sqrt{2}\sigma}{\delta}\right)^{m+i\gamma}\exp(in\theta)\times$$

$$\times\frac{\Gamma\left(\dfrac{n+m+i\gamma+2}{2}\right)\left(\dfrac{\rho^2}{2\omega_1^2}\right)^{n/2}}{n!}\,{}_1F_1\left(\frac{n+m+i\gamma+2}{2},n+1,-\frac{\rho^2}{2\omega_1^2}\right), \qquad (4.41)$$

where $\omega_1 = \dfrac{f_0}{k\sigma}$ is the effective radius of the light field in the Fourier plane. The light fields (4.39) are orthogonal with each other

for different values of the topological charge n. Note that when $f_0 = k\sigma^2$ the real part of the argument Kummer function no longer depends on the distance z: $\omega(z) = \sigma = $ const, and the imaginary part of the argument continues to be dependent on the variable z:

$R(z) = \frac{1}{2} f_0 \tan\left(\frac{z}{f_0}\right)$. Therefore, the amplitude of the entire field (4.39)

will depend on z. Therefore, the HyG beams (4.39) are not modes of a parabolic gradient medium for any parameters and values of the numbers m, n, γ, except for the Gaussian beam ($m = n = \gamma = 0$).

4.3.2. Modes of parabolic gradient medium

Consider a parabolic gradient medium of the form (4.32) $n^2(r) = n_0^2(1 - \tau^2 r^2)$, where $\tau = \alpha/2$. We will seek a solution to the Helmholtz equation for a parabolic medium.

$$\left(\frac{\partial^2}{\partial r^2} + \frac{1}{r}\frac{\partial}{\partial r} + \frac{1}{r^2}\frac{\partial^2}{\partial \varphi^2} + \frac{\partial^2}{\partial z^2} + k_0^2 n^2(r)\right)E(r,\varphi,z) = 0 \qquad (4.42)$$

with indicator (4.31) in modal form:

$$E(r,\varphi,z) = r^P z^q \exp\left(-\frac{r^2}{\omega_2^2}\right)\exp(in\varphi)\exp(i\beta z)F\left(r^t z^l\right), \qquad (4.43)$$

where F is a function, and β is the mode propagation constant. It can be shown that the solution of equation (4.42) is a family of functions:

$$E(r,\varphi,z) = r^n \exp\left(-\frac{r^2}{\omega_2^2}\right)\exp(\pm in\varphi)\exp(i\beta z) \times$$

$$\times {}_1F_1\left[\frac{n+1}{2} + \frac{\omega_2^2}{8}\left(\beta^2 - k^2 n_0^2\right), n+1, \frac{2r^2}{\omega_2^2}\right], \qquad (4.44)$$

where ${}_1F_1(a,b,x)$ is as previoysly the Kummer function, $\omega_2 = \sqrt{2/(kn_0\tau)}$. Since the Kummer function has the following asymptotics for $\xi \to \infty$ [53]:

$$ {}_1F_1(a,b,\xi) = \frac{\Gamma(b)}{\Gamma(a)}\exp(\xi)\xi^{a-b}\left(1 + O(1/\xi)\right), \qquad (4.45)$$

where $O(x)$ tends to zero faster than x, then the function (4.44) will diverge as $r \to \infty$. In the solution set (4.44) there are non-divergent solutions. If the first parameter in the Kummer function is a negative integer, the Kummer function becomes equal to a polynomial, and a solution of (4.44) converges to zero $r \rightarrow \infty$. That is, provided:

$$\frac{n+1}{2} + \frac{\omega_2^2}{8}\left(\beta^2 - k^2 n_0^2\right) = -s, \qquad (4.46)$$

instead of (4.44) we obtain the known Laguerre–Gauss modes:

$$E(r,\varphi,z) = r^n \exp\left(-\frac{r^2}{\omega_2^2}\right)\exp(\pm in\varphi)\exp(i\beta z)\,_1F_1\left[-s, n+1, \frac{2r^2}{\omega_2^2}\right] =$$
$$= \frac{s!}{(n+1)_s} r^n \exp\left(-\frac{r^2}{\omega_2^2}\right)\exp(\pm in\varphi)\exp(i\beta z)L_s^n\left(\frac{2r^2}{\omega_2^2}\right), \qquad (4.47)$$

where

$$\beta = kn_0\sqrt{1 - \frac{2\tau}{kn_0}(2s+n+1)}. \qquad (4.48)$$

If the radius of the Gaussian beam of the LG mode (4.46) does not satisfy the condition $\omega_2 = [2/(kn_0 t)]^{1/2}$, then the LG beam will no longer be a mode. The propagation of the Laguerre–Gauss non-mode beam in a parabolic medium was considered in [217].

Thus, this section shows that there is a wide class of mode solutions of the Helmholtz equation in a cylindrical coordinate system for a parabolic medium, but only those solutions from this class that coincide with the Laguerre–Gauss modes will have finite energy (that is, will be physically realizable) .

4.3.3. Parabolic gradient microlens

The solution family (4.39) enters as a fundamental beam a Gaussian beam, the amplitude of which also periodically repeats when propagating in a parabolic gradient fiber (4.32). From (4.30) with $n = m = \gamma = 0$, we can get:

$$E_0(\rho,\theta,z) = (-i)\left(\frac{2k\sigma^2}{f_2}\right)\left(1 - \frac{ik\sigma^2}{f_1}\right)^{-1}\exp\left(-\frac{\rho^2}{2\omega^2(z)} - \frac{ik\rho^2}{2R_1(z)}\right), \qquad (4.49)$$

where

$$R_1^{-1}(z) = \left[\cos^2(z/f_0) + \left(\frac{f_0}{k\sigma^2}\right)^2 \sin^2(z/f_0) - 1\right] \times$$

$$\times \left\{ f_0 \tan(z/f_0)\left[\cos^2(z/f_0) + \right. \right. \tag{4.50}$$

$$\left. \left. + \left(\frac{f_0}{k\sigma^2}\right)^2 \sin^2(z/f_0)\right]\right\}.$$

From (4.49) and (4.50) it follows that at $f_0 = k\sigma^2$ the Gaussian beam propagates in the parabolic fiber without change, maintaining its diameter. If $f_0 \neq k\sigma^2$, then the radius of the Gaussian beam varies according to the formula:

$$\omega(z) = \sigma \left[\cos^2\left(\frac{z}{f_0}\right) + \left(\frac{f_0}{k\sigma^2}\right)^2 \sin^2\left(\frac{z}{f_0}\right)\right]^{1/2}, \tag{4.51}$$

from which it follows that the minimum radius of $\omega_1 = f_0/(k\sigma)$ (if $f_0 < k\sigma^2$) is achieved at a distance $L_1 = \pi f_0/2$ from the beginning ($z = 0$). The diameter of the Gaussian beam in the half-decay of intensity will be equal to:

$$\text{FWHM} = \left(\frac{\sqrt{\ln 2}}{\pi}\right) \frac{\lambda r_0 n_0}{\sigma \sqrt{n_0^2 - n_1^2}}. \tag{4.52}$$

Thus, one can consider a gradient parabolic (GP) lens as a segment of a parabolic fiber with a radius of r_0 and a length along the optical axis $L_1 = (\pi f_0)/2$. Such a GP lens will focus a flat Gaussian beam incident on its input surface with a radius of σ into a focal spot with a diameter (4.52) that forms near the exit plane of the lens (Fig. 4.11).

From (4.52) follows the expression for the numerical aperture of the GP lens ($n_1 = 1$):

$$\text{NA} = \frac{\sigma}{r_0} \sqrt{n_0^2 - 1}. \tag{4.53}$$

At $\sigma = r_0$ the numerical aperture (4.53) coincides with NA for the Mikaelian planar lens [211]. For example, when $n_0 = 1.5$, $n_1 = 1$, $\sigma = r_0$, we obtain the following values of the length of the GP lens $L_1 = 2.1 r_0$ and the focal spot diameter FWHM = 0.36λ. This is slightly larger than the diffraction limit in a medium with a refractive index of $n_0 = 1.5$: FWHM = 0.34λ.

However, since the derivation of (4.52) assumes that the lens has unlimited sizes along the radius and refractive index (4.31) drops to zero at infinity, then such a value of the focus width (FWHM = 0.36λ) cannot be implemented in practice when the real lens is limited to a radius r_0 at which $n(r_0) = 1$.

4.3.4. Binary parabolic lens

The gradient parabolic lens can be approximately replaced by a binary parabolic lens according to the rule schematically shown in Fig. 4.12.

According to Fig. 4.12 the radius of the GP lens r_0 is divided into N equal segments of length Δ: $r_0 = N\Delta$, then the radii of the beginning and the ends of these segments are equal: $r_p = p\Delta$, $p = 0,1,2,... N–1$.

The beginning of the p-th binary ring coincides with the radius r_p, and the end of the p-th binary ring $\overline{r}_p$ is found from the equation:

$$\left(r_{p+1}^2 - \overline{r}_p^2\right) + n_0\left(\overline{r}_p^2 - r_p^2\right) = 2n_0 \int\limits_{r_p}^{r_{p+1}} \sqrt{1 - \tau^2 r^2}\, rdr. \qquad (4.54)$$

From equation (4.54) we can get an explicit expression for the radius of the end of the p-th binary ring:

$$\overline{r}_p^2 = \frac{\left(n_0 r_p^2 - r_{p+1}^2\right)}{(n_0 - 1)} + \frac{8}{3(n_0 - 1)\tau^2}\left[\left(1 - (\tau r_p)^2\right)^{1/3} - \left(1 - (\tau r_{p+1})^2\right)^{1/3}\right], \qquad (4.55)$$

where $p = 0,1,2,...,N–1$.

We note that in (4.55) the radii of the jumps of the refractive index r_p can be chosen not equidistant and the minimum technologically possible size of the band can be taken into account.

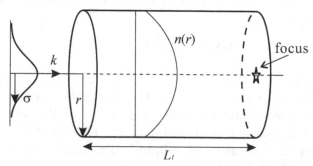

Fig. 4.11. Focussing scheme of a Gaussian beam using a GP lens.

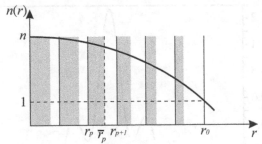

Fig. 4.12. Scheme of replacing the continuous parabolic dependence of the refractive index by the piecewise constant (binary) dependence.

4.3.5. Simulation results

Planar gradient parabolic lens

First, consider the two-dimensional case of a lens. Focussing of TE-polarized incident wave by FDTD method in the commercial program FullWave (Rsoft Design Group) was simulated. The grid of partitioning the counts along the optical axis Z and the transverse axis X was $\lambda/130$ (4.1 nm), the time step $\Delta(cT) = 2.8$ nm.

Figure 4.13 shows the refractive index in the lens in grayscale. Lens radius $r_0 = 1$ μm, $n_0 = 1.5$, $\alpha = 1.49$ μm^{-1}, $\lambda = 0.532$ μm. Since $\alpha = 2/f_0$, then $f_0 = 2/\alpha = 1.342$ mm, length of the lens $L_1 = \pi f_0/2 = \pi/\alpha = 2.1$ μm.

Figure 4.14 shows the intensity distribution of the electrical component at a distance of 10 nm behind the lens with a flat incident wave and a Gaussian beam with a radius of $\sigma = 1$ μm.

The width of the focal spot in the half intensity decay is FWHM = $0.388\lambda = 0.2$ μm in the case of a plane incident wave and FWHM = $0.5\lambda = 0.27$ μm in the case of a Gaussian beam. As can be seen from Fig. 4.14, the intensity maxima near the main focal spot are about 30%, which indicates a non-optimal length of the lens.

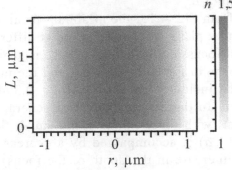

Fig. 4.13. The refractive index in a gradient binary lens in grayscale.

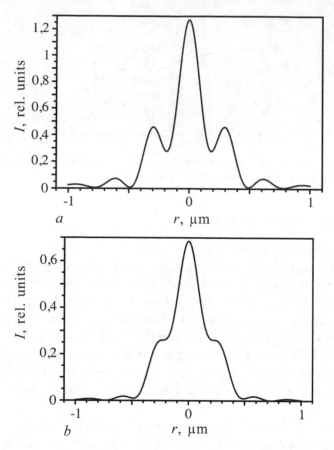

Fig. 4.14. The intensity distribution ($|E^2|$) in the focal plane at 10 nm behind the lens in the case of a plane incident wave (*a*) and a Gaussian beam (*b*).

Sharper focussing can be achieved by changing the length of the lens. Figure 4.15 shows the dependence of the width FWHM in wavelength of the light focussing of the light focussing by the lens on the lens with length L_1.

As can be seen from Fig. 4.15, the minimum focal spot width for a Gaussian beam and a plane wave is achieved at different lengths of the lens L_1. The best focussing of the plane wave is achieved with the value $L_1 = 1.56$ µm, with FWHM $= 0.375\lambda$, and the minimum focal spot with a Gaussian incident beam is FWHM $= 0.394\lambda$ with a lens length $L_1 = 1.73$ µm. In this result, there is a general pattern when focussing laser light: with other things being equal, the increase in side lobes (Fig. 4.14 *a*) is accompanied by a decrease in the focal spot width (and an increase in the depth of the focus).

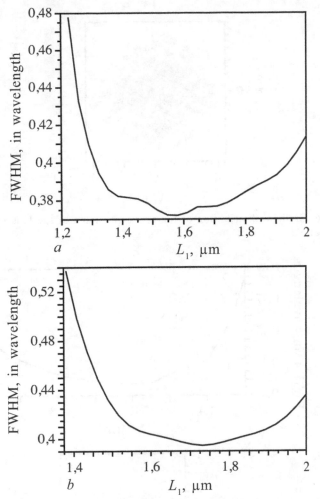

Fig. 4.15. The dependence of the focal spot width at 10 nm on the output plane of the lens on the lens length L for a plane wave (*a*) and a Gaussian beam (*b*) at the lens entrance.

In the three-dimensional case, a lens with a gradient refraction index in the *XY* plane depending on the radius *r* was simulated. The gradient distribution of the refractive index of a three-dimensional lens in the *XY* plane is shown in Fig. 4.16.

When simulating the three-dimensional case, a grid of $\lambda/40$ samples (13.3 nm) was used along all three axes. The propagation of linearly polarized light through a lens (the main component of the incident field E_y) was simulated; the incident wave is a flat Gaussian beam with a radius of $\sigma = 1$ μm. Figure 4.17 *a* shows the

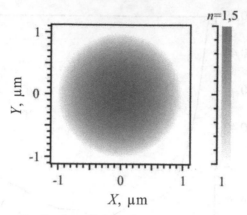

Fig. 4.16. Gradient distribution of the refractive index of the lens in the *XY* plane in grayscale.

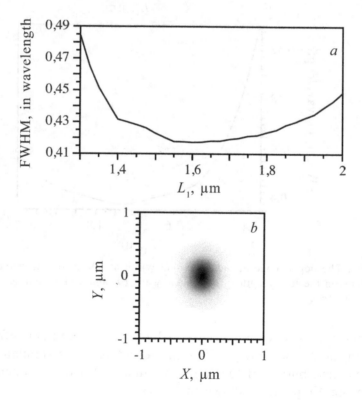

Fig. 4.17. Dependence of the focal spot diameter along the *X* axis at 10 nm on the output lens plane on the lens length L_1 for a linearly polarized Gaussian beam at the lens entrance with a radius $\sigma = 1$ μm (*a*) and intensity distribution (negative) in the lens focal spot at the optimum length $L_1 = 1.6$ μm (*b*).

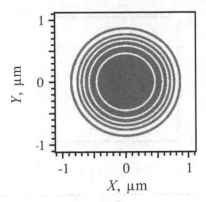

Fig. 4.18. The binary distribution of the refractive index of the lens in the *XY* plane. Dark colour corresponds to the refractive index $n = 1.5$, white $- n = 1$.

dependence of the FWHM focal spot width on the lens length L_1, measured through the centre of the focal spot along the *X* axis.

It can be seen that the minimum is reached with a lens length $L_1 = 1.6$ µm, and the width of the focus along the X axis is FWHM = 0.42λ. The intensity distribution ($|E^2|$) in the focal spot of the lens with these parameters is shown in Fig. 4.17 *b*. Due to linear polarization, the focal spot is broadened along the *Y* axis and amounts to FWHM = 0.70λ.

In practice, it is difficult to manufacture a lens with a gradient continuous profile of the refractive index. A binary microlens can be made according to the technology of manufacturing 3D photonic crystal waveguides or Bragg waveguides [218]. Figure 4.18 shows the binary distribution of the refractive index in the *XY* plane of the binary lens (4.55), which approximates a gradient parabolic lens.

Such a lens will focus light slightly worse than the variant with a continuous change in the refractive index (4.32). Figure 4.19 shows the intensity distribution in the focal plane of a binary parabolic lens and the cross section of this distribution through the centre along the *X* and *Y* axes. The lens length is optimal and is equal to $L_1 \approx 1.9$ µm (paraxial theory gives a slightly larger value $L_1 = 2.1$ µm). An incident linearly polarized field with a plane wave front has a Gaussian intensity distribution with a radius $\sigma = 1$ µm. The minimum size of the zone (the difference between adjacent radii of the refractive index jumps) of a binary lens is 35 nm. Simulation was carried out grid samples $\lambda/70$ (7.6 nm) in all three axes. The initial plane of polarization is parallel to the *ZY* plane.

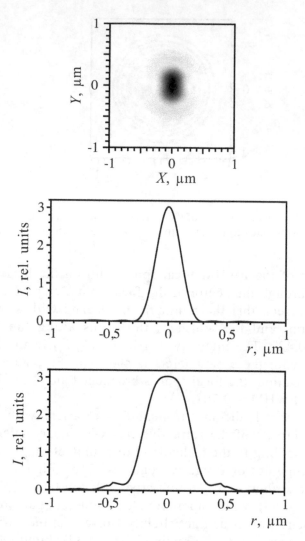

Fig. 4.19. The intensity distribution (negative) in the focal plane of the binary parabolic lens in Fig. 4.18 (*a*) and intensity cross sections through the centre of the focal spot along the *X* (*b*) and *Y* (*c*) axis.

The focal spot diameter by intensity half-decay (for Fig. 4.19) along the *X* axis is FWHM = 0.45λ, along the *Y* axis – FWHM = 0.78λ. The focal spot is elongated along the *Y* axis and has almost no side lobes. Recall that the paraxial theory gives equal to the diameter of the focal spot of FWHM = 0.36λ.

The following results were obtained in the section [62]. An expression for the complex amplitude of a family of paraxial hypergeometric laser beams propagating in a gradient parabolic fiber is obtained (Eq. (4.38)). A wide class of mode solutions of the Helmholtz equation in a cylindrical coordinate system for a gradient parabolic medium was found; these solutions are proportional to Kummer functions (equation (4.43)). It is shown that only those solutions from this class will have finite energy (that is, they will be physically realizable) that coincide with the Laguerre-Gauss modes (equation (4.46)). A segment of a certain length of a gradient parabolic fiber is considered as a parabolic lens, for which expressions are obtained for the numerical aperture and for the focus diameter by the intensity half-decay (equations (4.51), (4.52)). An FDTD simulation of focussing a linearly polarized Gaussian beam using a 3D gradient parabolic lens showed that the optimum lens length is less than the scalar theory predicts, and the smaller diameter of the elliptical focal spot is FWHM = 0.42λ (Fig. 4.17 *b*).

4.4. Optical elements to form the perfect vortex

In [219], the 'perfect' optical vortex (POV), which does not change its radius with a change in the magnitude of the topological charge, was first considered. Theoretically, such a vortex is described as a series of Bessel functions of the same order, but of different scales. In practice, such a series is replaced by a finite sum, which leads to errors in the formation of the 'perfect' optical vortex. In the experiment in [219] POV is formed using a phase optical element consisting of a finite set of concentric rings, the thickness of each of which should be as small as possible, and the phases in each ring depend linearly on the azimuth angle and are proportional to the topological charge. But, since the amplitude of the field in each ring must be different, the width of each ring, proportional to the amplitude, must also be different. This leads to conflicting requirements, on the one hand, the ring width should be minimal to match the width of the delta function, and, on the other hand, the width of each ring should be proportional to a predetermined amplitude. All this led to a low quality of the simulation results (Fig. 3 in [219]) and the experiment (Fig. 6 in [219]). In [220] a different procedure was used to produce the POV: using a conical axicon and a spiral phase plate.

However, the formula which, according to the authors, describes the amplitude of the POV formed by this procedure (equation (1) in [220]) describes the POV very approximately. In [221], a spatial light modulator is used, to which the phase mask is applied as a combination of an axicon and a spiral phase plate. With the help of this mask a Bessel–Gauss beam is formed, which passes through a lens that performs the Fourier transform. This approach is also based on an inaccurate formula describing diffraction on a screw axicon. In [222] the formation of the POV with the help of the axicon is investigated, but it is shown that the POV radius can be changed by simply moving the axicon, which eliminates the need to use lenses with a different focal distance. In [223], instead of an axicon, it is proposed to form a POV using an amplitude-phase optical element, the transmission function of which approximates a function describing the amplitude of the *n*-th order Bessel mode. In [223], the authors most closely approached the optimal phase element proposed in this section.

In this section, we will obtain formulas describing the distribution of the intensity of POV, formed in different ways, and it will be shown that the intensity on the POV ring is weak, but depends on the magnitude of the topological charge. We compare the formation of the POV by focussing: 1) the amplitude–phase light field forming the Bessel mode, bounded in radius, 2) the phase light field (constant amplitude) described in [6], and 3) the phase field (amplitude constant) formed using a conical axicon and a spiral phase plate [220]. All these three methods differ from the one proposed in [219]. We will show that the optimal phase optical element proposed in [224] is the best candidate for the formation of POV.

4.4.1. Formation of a perfect optical vortex using an amplitude-phase optical element

The 'perfect' optical vortex [219] has a complex amplitude of the form:

$$E_0(\rho,\theta) = \delta(\rho - \rho_0)\exp(in\theta), \qquad (4.56)$$

where $\delta(x)$ is the Dirac delta function, (ρ, θ) are the polar coordinates in the Fourier plane of the spherical lens, n is an integer equal to the topological charge of the optical vortex. From (4.56) it can be seen that the radius of the infinitely thin ring ρ_0 does not depend on the topological charge n. It is possible to form a POV (4.56) using the ideal Bessel mode, whose amplitude in the initial plane located at the focal distance from the spherical lens, has the form:

$$F_0(r,\varphi) = J_n(\alpha r)\exp(in\varphi), \qquad (4.57)$$

where the dimensional parameter α specifies the scale of the n-th order Bessel function of the first kind of the n-th order $J_n(x)$.

This follows from the condition of the orthogonality of the Bessel functions on the whole real axis [225]:

$$\alpha \int_0^\infty J_m(\alpha r) J_m\left(\frac{k\rho r}{f}\right) r\,dr = \delta\left(\alpha - \frac{k\rho}{f}\right), \qquad (4.58)$$

where k is the wave number of monochromatic coherent light, f is the focal length of the lens.

From (4.58) we obtain the radius of the ring with the maximum intensity of the POV, which is defined in (4.56):

$$\rho_0 = \frac{\alpha f}{k}. \qquad (4.59)$$

When forming the field (4.57), two problems arise. First, the Bessel function is an alternating, and therefore for its optical formation it is necessary to combine an amplitude mask having a transmission proportional to the function $|J_n(\alpha_r)|$ and a phase mask having a transmittance equal to:

$$\operatorname{sgn} J_n(\alpha r) = \begin{cases} 1, J_n(\alpha r) > 0, \\ -1, J_n(\alpha r) < 0. \end{cases} \qquad (4.60)$$

Secondly, the Bessel mode in (4.57) should in practice be limited to a circular aperture of radius R. Both of these causes lead to a distortion of POV.

In this section, the formation of the POV with the help of a light field (4.57), but limited to a circular aperture of radius R, is considered:

$$F_1(r,\varphi) = \operatorname{circ}\left(\frac{r}{R}\right) J_n(\alpha r)\exp(in\varphi), \qquad (4.61)$$

where

$$\operatorname{circ}\left(\frac{r}{R}\right) = \begin{cases} 1, r \le R, \\ 0, r > R. \end{cases} \qquad (4.62)$$

The complex amplitude of the light field (4.61) in the Fourier plane of an ideal spherical lens with focal length f can be found using the Fourier transform in cylindrical coordinates:

$$E_1(\rho,\theta) = (-i)^{n+1}\left(\frac{k}{f}\right)e^{in\theta}\int_0^R J_n(\alpha r)J_n\left(\frac{kr\rho}{f}\right)rdr =$$

$$= (-i)^{n+1}\left(\frac{kR}{f}\right)e^{in\theta}\left[\frac{\alpha J_{n+1}(\alpha R)J_n(xR) - xJ_n(\alpha R)J_{n+1}(xR)}{\alpha^2 - x^2}\right], \tag{4.63}$$

where $x = k\rho/f$. Note that the shape of the field (4.63) is significantly different from the specified POV (4.56). But despite this, the radius of the ring with the maximum intensity will be the same as (4.59) and will not depend on the topological charge. Indeed, it follows from (4.59) that the ring with the maximum intensity of the POV must have a radius at which the denominator in (4.63) vanishes. But when $x = \alpha$ the numerator also vanishes. Indeterminacy zero to zero in (4.63) at $x = \alpha$ (in the bright ring POV with maximum intensity) can be opened using the reference integral [65]:

$$\int_0^R J_n^2(\alpha r)rdr = \frac{R^2}{2}\left[J_n^2(\alpha R) - J_{n-1}(\alpha R)J_{n+1}(\alpha R)\right]. \tag{4.64}$$

Then, instead of (4.63) in view of (4.64) and with $x = \alpha$ we get:

$$E_1\left(\rho = \alpha fk^{-1},\theta\right) = (-i)^{n+1}\left(\frac{kR^2}{2f}\right)e^{in\theta}\left[J_n^2(\alpha R) - J_{n-1}(\alpha R)J_{n+1}(\alpha R)\right]. \tag{4.65}$$

From (4.65) it can be seen that the maximum intensity on the POV ring with constant values of α and R will depend on the value of the topological charge n. Although for large values of αR, the intensity will almost not depend on n, since, using the asymptotics of the Bessel function, we obtain that

$$J_n^2(\alpha R) - J_{n-1}(\alpha R)J_{n+1}(\alpha R) \approx \left(\frac{2}{\pi\alpha R}\right) \times$$

$$\times\left[\cos^2(\alpha R - n\pi/2 - \pi/4) - \right.$$
$$\left. -\cos(\alpha R - (n+1)\pi/2 - \pi/4)\cos(\alpha R - (n-1)\pi/2 - \pi/4)\right] =$$

$$= \left(\frac{2}{\pi\alpha R}\right)\left[\cos^2(\alpha R - n\pi/2 - \pi/4) + \sin^2(\alpha R - n\pi/2 - \pi/4)\right] =$$

$$= \left(\frac{2}{\pi\alpha R}\right). \tag{4.66}$$

From (4.66) it follows that the intensity on the ring with the maximum intensity of POV will asymptotically tend for large values of αR to a value that does not depend on the magnitude of the topological charge:

$$I_1\left(\rho = \frac{\alpha f}{k}\right) = |E_1|^2 = \left(\frac{kR}{\pi\alpha f}\right)^2. \tag{4.67}$$

To further obtain accurate characteristics of POV we select a scale of the Bessel function (4.60), namely, assume that $\alpha R = \gamma_{n,v}$, wherein $\gamma_{n,v}$ is the v-th root of the Bessel function: $J_n(\gamma_{n,v}) = 0$. Then instead of (4.63), we get:

$$E_1(\rho,\theta) = (-i)^{n+1}\left(\frac{kR^2}{f}\right)e^{in\theta}\left[\frac{\gamma_{n,v}J_{n+1}(\gamma_{n,v})J_n(xR)}{\gamma_{n,v}^2 - (xR)^2}\right]. \tag{4.68}$$

From (4.68) it can be seen that the zeros of the amplitude function (dark intensity rings of the POV) will coincide with the zeros of the n-th order Bessel function, and their radii will be equal to:

$$\rho_{n,\mu} = \frac{\gamma_{n,\mu}f}{kR}, \quad \mu \neq v, \tag{4.69}$$

and when $\mu = v$ there will be a maximum of intensity (bright ring POV), equal to:

$$I_1\left(\rho = \frac{\gamma_{n,v}f}{kR}\right) = |E_1|^2 = I_0 J_{n+1}^4(\gamma_{n,v}), \tag{4.70}$$

where $I_0 = [kR^2/(2f)]^2$ is the intensity in the centre of the Airy disk, i.e. in the centre of the Fraunhofer diffraction pattern of the plane wave of unit amplitude on a circular aperture with a radius R. Equation (4.70) follows from (4.65) if $\alpha R = \gamma_{n,\nu}$. Indeed, in view of (4.65), using recurrence relations for the Bessel function with $\alpha R = \gamma_{n,\nu}$ instead of (4.68), we obtain:

$$E_1\left(\rho = \frac{\gamma_{n,\nu} f}{kR}\right) = (-i)^{n+1} e^{in\theta}\left(\frac{kR^2}{2f}\right)\left[-J_{n-1}\left(\gamma_{n,\nu}\right)J_{n+1}\left(\gamma_{n,\nu}\right)\right] =$$

$$= (-i)^{n+1} e^{in\theta}\left(\frac{kR^2}{2f}\right)J_{n+1}^2\left(\gamma_{n,\nu}\right).$$

(4.71)

Equation (4.71) immediately follows from (4.70). From (4.70) it can be seen that the maximum intensity of the POV depends on n and on the value of the root $\gamma_{n,\nu}$. But the roots should not be chosen arbitrarily, but rather to satisfy the condition of preserving the scale of the Bessel function α and the radius of the aperture R, that is, the condition:

$$\alpha R = \gamma_{n,\nu} = \gamma_{m,\mu}.$$

(4.72)

This means that for different topological charges n and m, we need to choose such roots of the Bessel function in (4.57) so that condition (4.72) is satisfied. In this case, the ring with the maximum intensity will retain its radius as the magnitude of the topological charge changes. Note that the Bessel functions with different numbers do not have exactly the same roots, but one can always find two close roots.

From (4.68) one can estimate the width of the ring with the maximum intensity. It is equal to the distance between two zeros of the n-th order Bessel function with numbers $\nu - 1$ and $\nu + 1$:

$$\Delta\rho_0 = \frac{\left(\gamma_{n,\nu+1} - \gamma_{n,\nu-1}\right)f}{kR} \approx \frac{2\pi f}{kR}.$$

(4.73)

And the width of the POV ring in terms of intensity half-decay is approximately two times less: FWHM $= \pi f/(kR)$.

Let us estimate the efficiency of the formation of the POV using the light field (4.61). If illuminate a circular aperture of radius R by a plane wave with unit intensity, the power of light incident on the optical element with the transmission (4.6) is proportional to the area of a circle: $W_0 = \pi R^2$. The power of light transmitted from the optical element with transmission (4.61) is proportional to the right side of (4.64) multiplied by 2π, and in view of (4.66) for large αR is $W_1 = 2R/\alpha$. Therefore, the efficiency of POV formation via (4.61) will be no greater than $\eta = W_1/W_0 = 2/(\pi \alpha R)$. From this formula it follows that the efficiency decreases with increasing aperture radius.

4.4.2. Formation of a perfect optical vortex using the optimal phase optical element

By optimal we mean an optical element that directs the largest part of the light energy to a ring of a given radius. Work [224] described such an optical element, the transmission of which is equal to:

$$F_2(r,\varphi) = \text{circ}\left(\frac{r}{R}\right) \text{sgn} \, J_n(\alpha r) \exp(in\varphi). \tag{4.74}$$

In (4.74), the sign and aperture functions are defined in (4.61) and (4.61). The amplitude of the field at the focus of the spherical lens, formed by an optical element with transmission (4.74), but with $n = 0$, was obtained in [224]. Below we obtain an expression for the amplitude for any integer n. Let us proceed from the fact that a circle of radius R can be divided into N rings with radii r_m in which the sign of the function (4.60) is changed:

$$r_m = \frac{\gamma_{n,m}}{\alpha}, \, m = 1, 2, ..., N, \quad r_N = R. \tag{4.75}$$

The complex field amplitude at the focus of a lens formed by an optical element with the transmission function (4.74) can be expressed as the sum of the contributions from each such ring ($r_0 = 0$):

$$E_2(\rho,\theta) = (-i)^{n+1}\left(\frac{k}{f}\right)e^{in\theta}\sum_{m=0}^{N-1}(-1)^m \int_{r_m}^{r_{m+1}} J_n\left(\frac{k\rho r}{f}\right)rdr. \tag{4.75}$$

If we put in (4.76) $\rho = \alpha f/k$, then the argument of the Bessel function in (4.76) will not depend on the physical parameters f and k and will be equal to αr. Assume (without the loss of generality) scale parameter α such that Bessel function at the edge of the aperture was zero, i.e. $\alpha R = \gamma_n$, N.

The magnitude of the integrals in (4.76) for rings for which m is even, will be positive and the factor $(-1)^m$ will also be positive, and for rings for which m is odd, the value of the integrals in (4.76) will be negative, and the factor $(-1)^m$ will also be negative. That is, when $\rho_0 = \gamma_{n,N} f/(kR)$ sums all terms are positive, and the contribution to the field is the maximum at the ring radius. That is, the phase optical element (4.73) is indeed optimal from the point of view of forming the maximum intensity on a ring of a given radius $\rho_0 = \gamma_{n,N} f/(kR)$. To ensure that the radius of the POV ring does not depend on the topological charge, it is necessary to choose close roots of the Bessel function, similarly to (4.72): $\gamma_{n,N} = \gamma_{m,M}$. The above remarks about (4.76) mean that the intensity on the POV ring is equal to the following expression ($\alpha R = \gamma_{n,N}$):

$$I_2\left(\rho = \frac{\alpha f}{k}\right) = |E_2|^2 = \left(\frac{k}{f}\right)^2 \left(\int_0^R |J_n(\alpha r)| r \, dr\right)^2. \qquad (4.77)$$

Next, we obtain an explicit form of the intensity at the focus of the spherical lens for the initial field (4.74). In each ring bounded by radii (r_m, r_{m+1}) from (4.75), the radial transmission of the function (4.74) will be constant and change the sign from ring to ring. Therefore, the Fraunhofer diffraction of a plane wave on each ring of the field (4.74) can be described by the following complex amplitude [226]:

$$E_m(\rho,\theta) = (-i)^{n+1}\left(\frac{k}{f}\right)e^{in\theta}\int_{r_m}^{r_{m+1}} J_n\left(\frac{k\rho r}{f}\right)r\,dr = (-i)^{n+1}\left(\frac{1}{n!(n+2)}\right)e^{in\theta}\times$$

$$\times\left[\left(\frac{kr_{m+1}^2}{f}\right)x_{m+1}^n {}_1F_2\left(\frac{n+2}{2},\frac{n+4}{2}, n+1,-x_{m+1}^2\right) - \right. \qquad (4.78)$$

$$\left. -\left(\frac{kr_m^2}{f}\right)x_m^n {}_1F_2\left(\frac{n+2}{2},\frac{n+4}{2}, n+1,-x_m^2\right)\right],$$

where $m = 1, 2,.., N - 1$, $x_m = k\rho r_m/(2f) = k\rho\gamma_{n,m}/(2\alpha f)$, ${}_1F_2(a, b, c, x)$ is the hypergeometric function [65]. The total light field in the focus of the spherical lens from all ring apertures with radii (4.75) for the original field (4.74) is obtained by summing all the contributions (4.78) similarly to (4.76):

$$E_2(\rho,\theta) = -(-i)^{n+1} \left[\frac{k}{n!(n+2)\alpha^2 f} \right] e^{in\theta} \times$$

$$\times \left[2\sum_{m=1}^{N-1} (-1)^m \gamma_{n,m}^2 x_m^n {}_1F_2\left(\frac{n+2}{2}, \frac{n+4}{2}, n+1, -x_m^2 \right) - \right. \tag{4.79}$$

$$\left. -(-1)^N \gamma_{n,N}^2 x_N^n {}_1F_2\left(\frac{n+2}{2}, \frac{n+4}{2}, n+1, -x_N^2 \right) \right],$$

where $x_N = k\rho R/(2f)$. Substituting in (4.79) of the radius $r_0 = \gamma_{n,N} f/(kR)$ leads to the independent expression of the second square brackets on the physical parameters (α, k, f), as all the arguments will be proportional to the roots of the Bessel functions:

$$x_m = \frac{k\rho r_m}{2f} = \frac{k\rho\gamma_{n,m}}{2\alpha f} = \frac{\gamma_{n,m}}{2}. \tag{4.80}$$

Therefore, from (4.79) it is possible to obtain the dependence of the maximum intensity on the POV ring on these parameters:

$$I_2\left(\rho = \frac{\alpha f}{k} \right) = |E_2|^2 \sim \left(\frac{k}{\alpha^2 f} \right)^2. \tag{4.81}$$

Equation (4.81) is equal to (4.77), considering that $\alpha R = \gamma_{n,N}$. Note that the values of the arguments of the hypergeometric functions in each term (4.70) on the ring with maximum intensity of the POV depend on the roots of the Bessel function of the n-th order ($x_m = \gamma_{n,m}/2$). Therefore, if the topological charge of the POV is changed, the roots of the Bessel function and the values of the arguments will change. This will lead to a change in the amplitude modulus (4.79), and therefore to the change in intensity on the ring POV. The magnitude of this change can only be estimated using simulation. Since the optical element (4.74) is a phase element, the aperture radius and the POV radiu do not change when the topological charge n change. The change rate (4.77) of the field (4.79) on the ring of of this radius $r_0 = \gamma_{n,N} f/(kR)$ means the redistribution of energy between the POV ring and the side lobes. Note that the dependence of the intensity on the POV ring on the number n also follows directly from (4.77), since for the Bessel functions of different order the integral will have a different value.

4.4.3. Formation of a perfect optical vortex using a conical axicon

In [220], the POV is formed using a conical axicon and a spiral phase plate. In [220], only the experimental part is presented and there is no theory. Therefore, we will fill this gap. Instead of (4.61) and (4.74) in this section, we consider an optical element with a complex transmission function in the form:

$$F_3(r,\varphi) = \text{circ}\left(\frac{r}{R}\right)\exp(i\alpha r + in\varphi). \tag{4.82}$$

The optical element (4.82) was first considered in [227] as an element forming light tubes. Here α is not longer the scale parameter of the Bessel function, and the parameter of the axicon associated with the half-angle ψ at the top of the conical wave that formed these axicon, is connected by the expression: $\alpha = k \sin \psi$, although the axicon $\exp(i\alpha r)$ approximately forms the zero-order Bessel function with a scale of just α. This follows, for example, from the following equality [225]:

$$\exp(i\alpha r) = \sum_{m=-\infty}^{\infty} i^m J_m(\alpha r). \tag{4.83}$$

In [220] it is assumed that the light field (4.82) will form a POV at the focus of the spherical lens and its amplitude distribution is described by the function:

$$E_3(\rho,\theta) \sim \exp\left[-\frac{(\rho-\rho_0)^2}{\Delta\rho^2}\right]\exp(in\theta). \tag{4.84}$$

This is a strong simplification, and in fact the complex amplitude of the POV in the focus of a spherical lens is described as the Fourier transform of function (4.82) and looks much more complicated than function (4.84). We shall show it. The Fraunhofer diffraction of the light field (4.82) is described by the following expression [227]:

$$E_3(\rho,\theta) = (-i)^{n+1}\left(\frac{k}{f}\right)e^{in\theta}\int_0^R \exp(i\alpha r)J_n\left(\frac{k\rho r}{f}\right)r\,dr. \tag{4.85}$$

From (4.85) it can be seen that when the radius value is (4.59) (the radius of the ring with the maximum intensity of POV) in the integral, the scales of the two functions of the factors are aligned. Then we get:

$$E_3\left(\rho = \frac{\alpha f}{k}, \theta\right) = (-i)^{n+1}\left(\frac{k}{f}\right)e^{in\theta}\int_0^R \exp(i\alpha r)J_n(\alpha r)rdr =$$

$$= (-i)^{n+1}\left(\frac{k}{f}\right)e^{in\theta}\sum_{m=-\infty}^{\infty} i^m \int_0^R J_m(\alpha r)J_n(\alpha r)rdr. \qquad (4.86)$$

The second equality in (4.86) is obtained taking into account (4.83). From (4.86) it can be concluded that for a POV formed in the focus of the lens with the help of an optical element (4.82), the intensity on the ring of radius (4.59) will be maximum, since the multiplier functions under the first integral in (4.86) are matched most closely on the scale. Therefore, the intersection areas separately between the positive and separately between the negative components of these functions will be larger. The field intensity (4.86) will be much larger than the intensity on the POV (4.67) ring formed by the amplitude-phase element (4.62), since the intensity (4.67) is the contribution of only one term from the series to (4.86) with $m = n$. The above reasoning is qualitative. Using numerical simulation, it is shown below that the radius of the ring with maximum intensity and its magnitude change slightly with a change in the topological charge.

On the other hand, the intensity on the POV ring (4.86) will be less than the intensity on the POV ring (4.77) formed by the optimal phase element (4.74), since:

$$I_3\left(\rho = \frac{\alpha f}{k}\right) = |E_3|^2 = \left(\frac{k}{f}\right)^2 \left|\int_0^R \exp(i\alpha r)J_n(\alpha r)rdr\right|^2 <$$

$$< \left(\frac{k}{f}\right)^2 \left|\int_0^R |J_n(\alpha r)|rdr\right|^2 = I_2\left(\rho = \frac{\alpha f}{k}\right). \qquad (4.87)$$

Expression (4.87) shows that the intensity I_3 depends on the order n of the Bessel function at constant α, f, R. Since both optical elements (4.74) and (4.82) are phase and with the same radius R, and the intensity on the rings with the same radius (4.58) is different, as follows from (4.87), this means that the width of the POV ring (4.86) is greater than the width of the ring POV (4.76).

Physically, this can be explained as follows. The light ring (4.76) is formed by an optical element (4.74), which is partially a binary phase ring grating, forming mainly +1 and −1 diffraction orders, which both contribute to each point of the POV ring (4.76). That is, the entire circular aperture of radius R contributes to the formation of the ring, and therefore its width is approximately equal, like the width of the ring (4.63), to expression (4.73) or to the diffraction limit FWHM = $\pi f/(kR)$. And the POV light ring (4.86) is formed by the axicon (4.82), and only half of the aperture from 0 to R contributes to each point of the ring. Therefore, the width of the ring (4.86) is about 2 times wider than the rings (4.76) and (4.63). Simulation confirms this.

To obtain an explicit expression for the amplitude of the field (4.85), we expand the exponent under the integral in (4.85) into a Taylor series and obtain:

$$E_3\left(\rho,\theta\right)=(-i)^{n+1}\left(\frac{k}{f}\right)e^{in\theta}\sum_{m=0}^{\infty}\frac{(i\alpha)^m}{m!}\int_0^R r^m J_n\left(\frac{k\rho r}{f}\right)rdr. \qquad (4.88)$$

The integrals under the sum sign in (4.88) can be found in the reference book [65], therefore instead of (4.88) one can write down [227]:

$$E_3\left(\rho,\theta\right)=\frac{(-i)^{n+1}}{n!}\left(\frac{kR^2}{f}\right)e^{in\theta}\,x^n\sum_{m=0}^{\infty}\frac{(i\alpha R)^m}{(m+n+2)m!}\times$$

$$\times{}_1F_2\left(\frac{m+n+2}{2},\frac{m+n+4}{2},n+1,-x^2\right), \qquad (4.89)$$

where $x = k\rho R/(2f)$, ${}_1F_2(a,\ b,\ c,\ x)$ is the hypergeometric function [65]. If the POV formed by the optimal element (4.74) was described by the finite sum of hypergeometric functions (4.79), then the POV formed by the axicon (4.82) is described by a series of such functions (4.89) and has nothing in common with the expression (4.84) from [220]. The light ring with the maximum intensity of the POV has a radius (4.59). To preserve the POV rings radius when changing the topological charge n we should select as previously close roots $\alpha R = \gamma_{n,N} \approx \gamma_{m,M}$. The argument x in (4.89) on the ring with the maximum intensity will be $x = \alpha R/2 = \gamma_{n,N}/2$. Therefore, the argument is x, and the whole expression under the sign of the sum in (4.89) will not depend on the physical parameters of the $k,\ f,\ R$. Consequently, similarly to (4.81), one can write an expression for the dependence of the maximum intensity on the POV ring (4.87) on physical parameters:

$$I_3\left(\rho = \frac{\alpha f}{k}\right) = |E_3|^2 \sim \left(\frac{kR^2}{f}\right)^2. \tag{4.90}$$

The dependence (4.90) is the same for all three considered elements (4.61), (4.74) and (4.82).

4.4.4. Simulation results

This section presents the results of simulating the POV formation using the three optical elements discussed in the previous sections. These results are consistent with the predictions of the theory described in the previous three sections. Simulation was performed with the following parameters: the radiation wavelength $\lambda = 532$ nm, the radius of the circular aperture of $R = 20\lambda$, the focal length of perfect spherical lens $f = 100\lambda$, scale parameter of the Bessel function α was chosen such that the following equality holds: $\alpha R = \gamma_{1.20} = 63.6114$, where $\gamma_{1.20}$ is the 20th root of the Bessel function ($v = 20$) of the first order ($n = 1$). The formation of the POV was compared for two different topological charges $n = 1$ and $n = 14$. The remaining simulation parameters were preserved. In this case, for the 14th order Bessel function ($n = 14$) the 14th root ($v = 14$) was chosen, since $\gamma_{14.14} \approx \gamma_{1.20} = 63.6114$. Figure 4.20 shows the modules of the two Bessel functions $|J_1(\gamma_{1.20}x/R)|$ and $|J_{14}(\gamma_{14.14}x/R)|$, limited

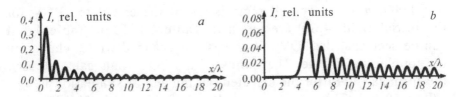

Fig. 4.20. Bessel function modules $|J_1(\gamma_{1.20}x/R)|$ (*a*) and $|J_{14}(\gamma_{14.14}x/R)|$ (*b*) limited to radius R.

Table 4.1. Comparison of the parameters of the POV formed by the initial light field (4.61) with different topological charges n

Topological charge	$n = 1$	$n = 14$
Radius of the ring with maximum intensity, ρ_0, λ	50.781563	50.781563
Maximum intensity in relative units, I_{max}	0.0157968	0.0150522
Ring thickness by half intensity decay, FWHM, λ	2.244489	2.244489

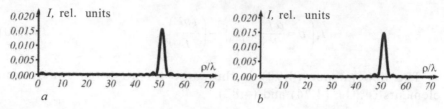

Fig. 4.21. The intensity distribution of the POV with $n = 1$ (*a*) and $n = 14$ (*b*) for the initial light field (4.61). On the ordinate axis are the relative units, and on the *x*-axis the radius in wavelengths

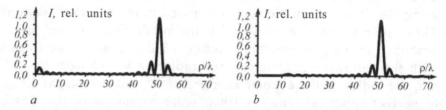

Fig. 4.22. The intensity distribution of the POV with $n = 1$ (*a*) and $n = 14$ (*b*) for the initial light field (4.74). On the ordinate axis are the relative units, and on the *x*-axis the radius in wavelengths

by the radius R. From Fig. 4.20 it is clear that for $x = R$ both Bessel functions have zero.

Figure 4.21 shows the intensity distribution of the POV in the Fourier plane of a spherical lens, obtained at the initial light field with a complex amplitude (4.61) with different topological charges $n = 1$ and $n = 14$. The characteristics of the calculated POV for the initial field (4.61) are given in Table 4.1. From Table 4.1 it can be seen that the POV radius has not changed with a change in the topological charge. The radius of the POV rings calculated by the formula (4.58) for the selected parameters of simulation is $\rho_0 = af/k \approx 50.62\lambda$. This ring radius value differs by only 3% from the radius value in Table 4.1. The maximum intensity of the POV decreased by only 5% with an increase in the topological vortex charge by almost an order of magnitude.

Note that according to formula (4.67) with the selected simulation parameters, the maximum intensity in Fig. 4.21 must be equal to $I_1(\rho_0) = [kR/(\pi af)]^2 \approx 0.015816$. This number is consistent with the intensity value from Table 4.1 (a difference of 0.1%). And since the radius of the ring has not changed and the radius of the aperture R of the optical element has not changed either, the width of the ring should not change either. Table 4.1 shows that the width of the ring does not change with a change in the topological charge of the optical vortex. According to (4.74) the ring width for the selected simulation parameters should be equal to FWHM = $5/2\lambda$. This value is 11% different from the width of the ring from Table 4.1.

Let us now consider the formation of the POV with the help of the optimal phase element (4.74). Figure 4.22 shows the intensity distribution of POV with $n = 1$ (*a*) and $n = 14$ (*b*) for the initial light field (4.74). Table 4.2 shows the calculated parameters of the POV in Fig. 4.22. From Table 4.2 it can be seen that the radius of the POV ring has become slightly smaller than in Fig. 4.22 (less than 0.3%). The radius did not change when the topological charge increased 14 times. The intensity on the ring is almost 100 times greater than the intensity for the POV in Fig. 4.21. This intensity can be calculated by the formula (4.64). Note that with an increase in the topological charge of 14 times, the intensity on the ring decreased by only 2%. The thickness of the ring has become less by about 14%, compared with the width of the ring in Fig. 4.21. The ring thickness is maintained when the topological charge of the optical vortex changes. Figure 4.22 shows that the side lobes have increased.

Next, we consider the formation of the POV with the help of the vortex axicon (4.82). Figure 4.23 shows the distribution of the intensity of POV with $n = 1$ (*a*) and $n = 14$ (*b*) for the initial light

Table 4.2. Comparison of the parameters of the POV, formed by the optimal phase element (4.73) with different topological charges n

Topological charge	$n = 1$	$n = 14$
Radius of the ring with maximum intensity, ρ_0, λ	50.641283	50.641283
Maximum intensity in relative units, I_{max}	1.140685	1.1181689
Ring thickness by half intensity decay, FWHM, λ	1.9639279	1.9639279

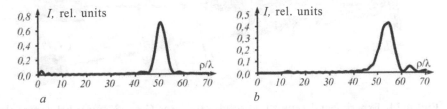

Fig. 4.23. The distribution of the intensity of POV with $n = 1$ (*a*) and $n = 14$ (*b*) for the initial light field (4.82). On the ordinate axis are the relative units, and on the *x*-axis the radius in wavelengths.

Table 4.3. Comparison of the parameters of the POV formed by the vortex axicon (4.82) with different topological charges n

Topological charge	$n = 1$	$n = 14$
Radius of the ring with maximum intensity, ρ_0, λ	50.501002	54.849699
Maximum intensity in relative units, I_{max}	0.7070332	0.4249419
Ring thickness by half intensity decay, FWHM, λ	4.9098196	6.5931864

field (4.82), and in Table 4.3 introduced the calculated parameters of this POV.

Figure 4.23 and Table 4.3 shows that the width of the POV ring formed by the vortex axial (4.82) at the focus of the spherical lens is approximately 2.5 times greater than the width of the ring in Fig. 4.21.

And, moreover, with an increase in the topological charge of the vortex by 14 times, the width of the ring increases by 1.3 times.

An increase in the width of the ring (Fig. 4.23) with an increase in the number n leads to a decrease in intensity on this ring. From Tablw 4.3 it can be seen that the maximum intensity on the POV ring (Fig. 4.22) decreases 1.7 times as the number n increases 14 times. And even the radius of the ring with the maximum intensity increases with 8%.

Thus, the simulations showed that of the three considered variants of the formation of the POV the best is the optimal phase element (4.74) since in this case it is formed by the narrowest band of light (FWHM = $1.96\lambda = 0.39\lambda f/R$) with maximum intensity on the ring,

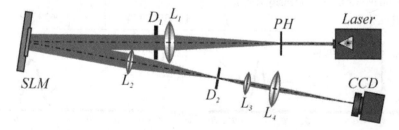

Fig. 4.24. Experimental setup: L – solid-state laser ($\lambda = 532$ nm), PH – pinhole (hole size 40 μm), L_1, L_2, L_3 and L_4 – lenses with focal lengths ($f_1 = 250$ mm, $f_2 = 350$ mm, $f_3 = 150$ mm, $f_4 = 500$ mm), D_1 and D_2 – diaphragm, SLM – spatial light modulator PLUTO VIS, CCD - LOMO TC -1000 video camera.

1.6 times greater than the intensity on the POV ring formed by the vortex axicon (4.82).

4.4.5. Experiment

For an experimental study of the elements forming the POV, we used the optical scheme shown in Fig. 4.24. A solid-state laser Laser (λ = 532 nm), which forms the fundamental Gaussian beam, was chosen as the radiation source. The laser beam, expanded and collimated with the help of a system consisting of PH pinhole (hole size 40 μm) and lens L_1 (f_1 = 250 mm), fell on the display of the SLM modulator (PLUTO VIS, resolution 1920 × 1080 pixels, pixel size 8 um). Aperture D_1 was used to separate the central bright spot from the bright and dark rings surrounding it, arising from the diffraction on the pinhole. Then, using a lens system L_2 (f_2 = 350 mm) and L_3 (f_3 = 150 mm) and a diaphragm D_2, we carried out spatial filtering of the phase modulated laser beam reflected from the modulator display. Using the L_4 lens (f_4 = 500 mm), the laser beam was focused on the CCD LOMO TC 1000 video camera matrix (pixel size 1.67 × 1.67 microns). To form the POV, we used the phase patterns shown in Fig. 4.25, which were displayed on the display of the light modulator. In order to space apart the unmodulated beam reflected from the modulator and the phase modulated beam, a linear phase mask was

Table 4.4. Comparison of the parameters of POV, formed by the optimal phase element with the topological charge $n = 1$ and $n = 14$

Topological charge	$n = 1$	$n = 14$
Ring radius with maximum intensity, μm	1491.0 ± 2.0	1496.5 ± 2.0
Maximum intensity, rel. units	156.0 ± 0.5	151.0 ± 0.5
Ring thickness by intensity half-decay, μm	70.0 ± 2.0	73.0 ± 2.0

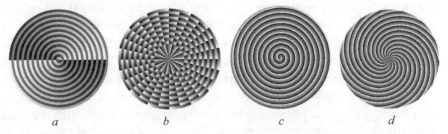

a	*b*	*c*	*d*

Fig. 4.25. Phase patterns of optical elements to form a POV with a topological charge $n = 1$ (*a, c*) and $n = 14$ (*b, d*). In the figures (*a, b*) the optimal phase elements are presented, in the figures (*c, d*) – vortex axicons.

Table 4.5. Comparison of the parameters of the POV formed by the vortex axicon with the topological charge $n = 1$ and $n = 14$

Topological charge	$n = 1$	$n = 14$
Ring radius with maximum intensity, µm	1498.0 ± 2.0	1655.0 ± 2.0
Maximum intensity, rel. units	96.0 ± 0.5	43.0 ± 0.5
Ring thickness by intensity half-decay, µm	158.0 ± 2.0	206.0 ± 2.0

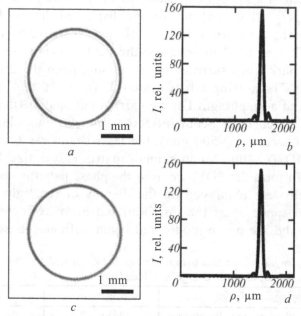

Fig. 4.26. The intensity distribution of the POV (negative, left column) and the corresponding cross sections from the centre of the image to the edge (right column) when using the optimal phase element with a topological charge $n = 1$ (*a, b*) and $n = 14$ (*c, d*).

additionally superimposed on the initial phase pattern of the element.

Figure 4.26 shows the intensity distributions formed at the focus of the lens L_4 using phase masks corresponding to the optimal phase elements with the topological charges 1 and 14. The values of the parameters of the formed POV are presented in Table 4.4.

Figure 4.27 shows the intensity distribution formed in the focus of the lens L_4 using the phase masks corresponding to the vortex axicons with topological charges 1 and 14. The values of the parameters of the formed POV are presented in Table. 4.5.

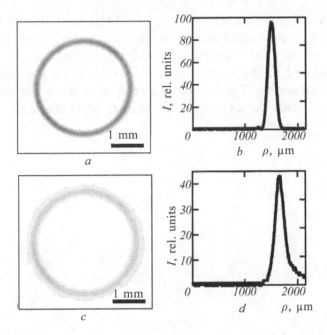

Fig. 4.27. The intensity distribution of the POV (negative, left column) and the corresponding cross sections from the centre of the image to the edge (right column) in the case of using a vortex axicon with topological charge $n = 1$ (a, b) and $n = 14$ (c, d).

Thus, from the analysis of experimental measurements of the POV parameters, it follows that the relative values of the parameters of the generated beams are in good agreement with the presented simulation results.

In this section we obtained exact analytical expressions [228], which describe the complex amplitude of the light field of an perfect optical vortex formed by three different optical elements: amplitude–phase with transmittance, proportional to the complex function describing the Bessel mode, the optimum phase element and the vortex axicon. It is shown that the use of any of these three optical elements leads to the formation of bright rings with the same radius, which weakly depends on the topological charge of the optical vortex. The intensity of light on the ring is greater (other things being equal) for the optimum phase element. For the optimum phase element, the light intensity on the ring weakly depends on the magnitude of the topological charge (Table 4.4), and for the vortex axicon both in the simulation (Table 4.3) and in the experiment (Table 4.5), the intensity on the ring decreases almost twice as the topological charge increases from 1 to 14. The width of the light ring formed by the vortex axicon is approximately 2 times more than the width of the other two rings. Thus, the optimal element (4.74), first considered in [6], is the best candidate for the formation of an ideal optical vortex. The simulation results confirm the theoretical conclusions, and the experimental results are consistent with the theory and simulation results.

Vector vortex beams

5.1. Hankel scalar nonparaxial beams

In optics there are laser beams described by scalar complex amplitudes, which are exact solutions of the nonparaxial Helmholtz equation. These are well-known plane and spherical waves [59], as well as the recently obtained Bessel modes [1, 125], Mathieu beams [4, 116], parabolic laser beams [229], Hankel–Bessel beams [146], and asymmetric Bessel modes [88,124].

In [230] expressions were obtained for the projections of the strength of the electric (E-vector) and magnetic (H-vector) electromagnetic wave fields. All projections of the E- and H-vectors are expressed through three integrals, which are expansions in the angular spectrum of plane waves. Based on the projections of the electromagnetic field, expressions are obtained for the Poynting vector and the density of the orbital angular momentum (OAM). We note that closed analytical expressions for OAM have so far obtained only for the paraxial Laguerre–Gauss [9] and Bessel vector beams [131].

In this section we consider new nonparaxial vortex beams where the complex amplitude is described by the Hankel function of half–integer order, therefore, these beams are called Hankel beams. The lowest order (zero) Hankel beam coincides with a spherical wave. Three types of such beams are considered, which are obtained from each other by taking derivatives. The considered laser beams can be formed using a liquid-crystal microdisplay and used for optical capture and rotation of dielectric microobjects.

It is known that the complex amplitude of the nonparaxial stationary light field $E(x,y,z)$ satisfying the Helmholtz equation can be represented as the angular spectrum of plane waves

$$E(x,y,z) = \iint_{\mathbb{R}^2} A(\xi,\eta) \exp\left[ik(\xi x + \eta y) + ikz\sqrt{1-\xi^2-\eta^2}\right] d\xi d\eta, \qquad (5.1)$$

where k is the wave number of monochromatic light, $A(\xi, \eta)$ is the complex amplitude of the angular spectrum of the plane waves. In polar coordinates (r, φ) expression (5.1) takes the form:

$$E(r,\varphi,z) = \iint_{\mathbb{R}^2} A(\rho,\theta) \exp\left[ikr\rho\cos(\theta-\varphi) + ikz\sqrt{1-\rho^2}\right] \rho d\rho d\theta, \qquad (5.2)$$

where (ρ, θ) are the polar coordinates in the Fourier plane. If the amplitude of the spectrum of plane waves is chosen in the form

$$A(\rho,\theta) = \frac{\rho^n}{\sqrt{\rho^2-1}} \exp(in\theta), \qquad (5.3)$$

then instead of (5.2) we can write:

$$E_{1,n}(r,\varphi,z) = 2\pi i^n \exp(in\varphi) \times$$
$$\times \int_0^\infty \frac{\rho^{n+1}}{\sqrt{\rho^2-1}} \exp\left(-kz\sqrt{\rho^2-1}\right) J_n(kr\rho) d\rho, \qquad (5.4)$$

where $J_n(x)$ is an nth order Bessel function of the first kind. The integral in (5.4) is a reference integral [65], and instead of (5.4) we can write:

$$E_{1,n}(r,\varphi,z) = i^{n-1}\pi\sqrt{\lambda}\, \frac{r^n}{\left(r^2+z^2\right)^{(2n+1)/4}} H_{n+1/2}^{(2)}\left(k\sqrt{r^2+z^2}\right) \exp(in\varphi), \qquad (5.5)$$

where $H_\nu^{(2)}(x) = J_\nu(x) - iY_\nu(x)$ is the Hankel function, $Y_\nu(x)$ is the Neumann function. The beams (5.5) are called Hankel beams of the 1st type (H1-beams). The Hankel beam (5.5) is a linear optical vortex $(x+iy)^n$ embedded in a generalized spherical wave $\psi_{n+1/2}(R) = H_{n+1/2}^{(2)}(kR) R^{-n-1/2}$, where $R = (r^2+z^2)^{1/2}$. The function depends only on the distance to the centre of the coordinate system $\psi(R)$, therefore it can be called a generalized spherical wave. Amplitude (5.5) has a special point when $r = z = 0$ in which it is equal to infinity. It can be shown that the amplitude of the H1-beam with $n = 0$ or $n = 1$ is related to the amplitude of the spherical wave.

Fundamental H1-beam (with $n = 0$) coincides with a spherical wave propagating from the origin:

$$E_{1,0} = -i\pi\sqrt{\lambda}\,\frac{H_{1/2}^{(2)}(kR)}{R^{1/2}} = 2\pi\frac{e^{-ikR}}{kR},\ R = \sqrt{r^2 + z^2}.\qquad(5.6)$$

With $n = 1$, we obtain the derivative of a spherical wave with a linear optical vortex 'embedded' on the axis:

$$E_{1,1} = \pi\sqrt{\lambda}\left(re^{i\varphi}\right)\frac{H_{3/2}^{(2)}(kR)}{R^{3/2}} = -2\pi\left(\frac{re^{i\varphi}}{kR}\right)\frac{d}{dR}\left(\frac{e^{-ikR}}{kR}\right).\qquad(5.7)$$

The Hankel beams of the 2nd type can be obtained by differentiating the complex amplitude (5.5) by the z coordinate:

$$E_{2,n}(r,\varphi,z) = 2\pi i^n k^{-1}\exp(in\varphi)\frac{\partial}{\partial z}\times$$

$$\times\left\{\int_0^\infty\frac{\rho^{n+1}}{\sqrt{\rho^2-1}}\exp\left(-kz\sqrt{\rho^2-1}\right)J_n(kr\rho)d\rho\right\} =\qquad(5.8)$$

$$= -2\pi i^n\exp(in\varphi)\int_0^\infty\rho^{n+1}\exp\left(-kz\sqrt{\rho^2-1}\right)J_n(kr\rho)d\rho.$$

The integral in (5.8) is also a reference integral [65], although it is possible to directly differentiate the right-hand side of the equality in (5.5) and use the recurrence relations for cylindrical functions. Instead of (5.8) we get:

$$E_{2,n}(r,\varphi,z) = i^{n+1}\pi\sqrt{\lambda}\,\frac{zr^n}{\left(\sqrt{r^2+z^2}\right)^{n+3/2}}H_{n+3/2}^{(2)}\left(k\sqrt{r^2+z^2}\right)\exp(in\varphi).\qquad(5.9)$$

The beams (5.9) are called the Hankel beams of the 2nd type (H2-beams). These beams, as well as the H1-beams, are structured as generalized spherical waves with linear optical vortices introduced. The only difference is that the function $\psi(R)$ has a slightly different form $\psi_{n+3/2}(R) = H_{n+3/2}^{(2)}(kR)R^{-n-3/2}$, and in the initial plane ($z = 0$) amplitude (5.9) is non-zero only at the origin.

Using the well-known approximate formulas for small and large values of the arguments of the Hankel functions

$$\begin{cases} H_\nu^{(2)}(x) \approx \dfrac{i}{\pi}\Gamma(\nu)\left(\dfrac{2}{x}\right)^\nu, 0 < x \ll 1, \\[4mm] H_\nu^{(2)}(x) \approx \sqrt{\dfrac{2}{\pi x}}\, i^{\nu+2}\exp(-ix), x \gg 1. \end{cases} \qquad (5.10)$$

Asymptotic dependences can be obtained for the H1- and H2-beams:

$$\begin{cases} |E_1| \sim r^{-1}, r \gg z, \\[4mm] |E_1| \sim \dfrac{r^n}{z^{2n+1}}, r \ll z, \end{cases} \qquad (5.11)$$

$$\begin{cases} |E_2| \sim r^{-2}, r \gg z, \\[4mm] |E_2| \sim \dfrac{r^n}{z^{2n+2}}, r \ll z. \end{cases} \qquad (5.12)$$

From (5.12) it can be seen that the intensity of both types of beams with a fixed z and with increasing r increases as r^n near the optical axis (this is the characteristic behaviour for the vortex beam), but at large r the intensity of the H2-beams decreases faster (proportional to r^{-4}) than the intensity of the H1-beams (decreases as for a spherical wave in proportion to r^{-2}). This means that different types of Hankel beams have different divergences. The divergence of both types of Hankel beams can be estimated for large z. It turns out to depend linearly on z. Indeed, the intensity of the H1-beams (5.5) is equal to:

$$I_1(r,z) = 2\pi^3 \frac{(kr)^{2n}}{(kD)^{2n+1}}\left[J_{n+1/2}^2(kD) + Y_{n+1/2}^2(kD)\right] \approx$$

$$\approx 2\pi^3 \frac{(kr)^{2n}}{(kD)^{2n+1}}\frac{2}{\pi kD} = \lambda^2 r^{2n} D^{-2n-2}, \qquad (5.13)$$

where $D = \sqrt{r^2 + z^2}$. In (5.13), in obtaining the approximate equality, the well-known asymptotics of the Bessel and Neumann functions were used. Differentiating (5.13) on the radial variable r and equating zero, we get:

$$\frac{\partial I_1}{\partial r} = 0 \Rightarrow r \approx z\sqrt{n}. \qquad (5.14)$$

From (5.14) it is seen that the radius of the maximum intensity of the vortex H1-beams ($n > 0$) linearly depends on the longitudinal coordinate z and is proportional to the square root of the topological charge of the optical vortex. Acting similarly for H2-beams, we find the intensity at large z:

$$I_2(r,z) = 2\pi^3 (kz)^2 (kr)^{2n} (kD)^{-2n-3} \left[J_{n+3/2}^2(kD) + Y_{n+3/2}^2(kD) \right] \approx$$
$$\approx 2\pi^3 \frac{(kz)^2 (kr)^{2n}}{(kD)^{2n+3}} \frac{2}{\pi kD} = \lambda^2 z^2 r^{2n} D^{-2n-4}, \tag{5.15}$$

and equate its derivative to zero, we get:

$$\frac{\partial I_2}{dr} = 0 \Rightarrow r = \sqrt{\frac{n}{2}} z. \tag{5.16}$$

From a comparison of (5.14) and (5.16) it can be seen that the divergence of the H2-beams is $\sqrt{2}$ times smaller than the divergence of the H1-beams. Approximately we can say that the H1-beams with $n = 1$ diverge in an angle of 45 degrees, and H2-beams – in an angle of 32 degrees. As the number n grows, the angles of divergence increase.

For any z, it is possible to accurately determine the dependence of the radius of maximum intensity for an H1-beam with $n = 1$. In this case, the intensity is

$$I_1(r,z) = \pi^2 \lambda r^2 D^{-3} \left[J_{3/2}^2(kD) + Y_{3/2}^2(kD) \right] =$$
$$= (2\pi)^2 k^2 r^2 \left[\frac{1}{(kD)^6} + \frac{1}{(kD)^4} \right]. \tag{5.17}$$

Equating to zero the derivative of intensity (5.17) with respect to the variable r, we get:

$$\frac{\partial I_1}{\partial r} = 0 \Rightarrow r = k^{-1} \sqrt{\sqrt{1 + (kz)^2 + (kz)^4} - 1}. \tag{5.18}$$

Figure 5.1 shows the intensity (a, d), phase (b, e) and the radial cross-section of intensity (c, f) for the H1-beam with topological charge $n = 1$ at distances $z = \lambda(a, b, c)$ and $z = 5\lambda(d, e, f)$. On the horizontal axis Fig. 5.1 shows the radial coordinate in wavelengths.

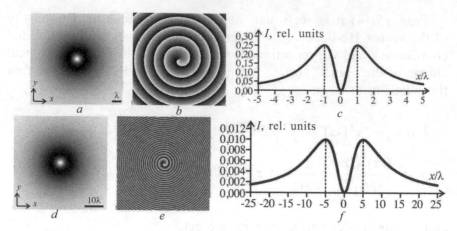

Fig. 5.1. Intensity (negative) (*a, d*), phase (*b, e*) and radial cross-section of intensity (*c, f*) for an H1-beam with a topological charge $n = 1$ at distances $z = \lambda(a, b, c)$ and $z = 5\lambda(d, e, f)$.

The vertical lines in Fig. 5.1 *c* and Fig. 5.1 *f* show the intensity maxima in the cross section, the radii of which can be accurately determined by the formula (5.18). For example, for the section in Fig. 5.1 *c* the radius of maximum intensity is $r = 0.994\lambda$. That is, not only in the far zone (5.14), but also in the near zone, the H1-beam divergence ($n = 1$) is equal to 45 degrees.

Note that both types of beams do not have side lobes. There are articles [84, 231, 232] in which the contrast of the side lobes of laser optical vortices is reduced by special methods in order to effectively use them for optical capture of micro-objects. In this case, there is no need for this, since the Hankel beams propagate without the side lobes.

Figure 5.2 shows the intensity (*a, d*), phase (*b, e*) and the radial cross-section of intensity (*c, f*) for the H1-beam (*a, b, c*) and H2-beam (*d, e, f*) with topological charges $n = 5$ at a distance $z = 5\lambda$. It can be seen from Figs. 5.2 *c* and 5.2 *f* that the divergence of the H1-beam is greater than that of the H2-beam with the same topological charges. The radius of maximum intensity in Fig. 5.2 is equal to 11λ, and in Fig. 5.2 *f* to 7.9λ.

Similarly, both from the amplitude of the H1-beam (5.5) using differentiation (5.8) we can obtain the H2-beam (5.9), also from the H2-beam by differentiation with respect to *z* we can receive Hankel type 3 beams (H3-beams):

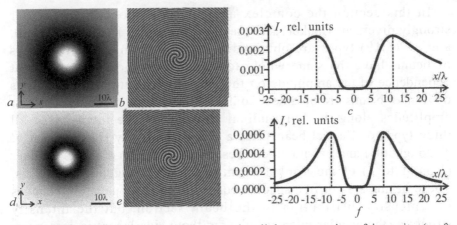

Fig. 5.2. Intensity (*a*, *d*), phase (*b*, *e*) and radial cross section of intensity (*c*, *f*) for the H1-beam (*a*, *b*, *c*) and H2-beam (*d*, *e*, *f*) with topological charges $n = 5$ at a distance $z = 5\lambda$.

$$E_{3,n}\left(r,\varphi,z\right) = i^{n-1}\pi\sqrt{\lambda}k^{-1}r^{n}\exp\left(in\varphi\right)\times$$

$$\times\left[\frac{kz^{2}H_{n+5/2}^{(2)}\left(k\sqrt{r^{2}+z^{2}}\right)}{\left(r^{2}+z^{2}\right)^{(2n+5)/4}} - \frac{H_{n+3/2}^{(2)}\left(k\sqrt{r^{2}+z^{2}}\right)}{\left(r^{2}+z^{2}\right)^{(2n+3)/4}}\right]. \tag{5.19}$$

Note that all Hankel beams (5.5), (5.9) and (5.19) have the same topological charge *n*, although the radial structure of the beams is described by the Hankel functions of different orders. This means that although the orbital angular momentum (OAM) of these beams is normalized per unit of power (or per photon), they are the same (it is equal to the topological charge), but the non-normalized OAMs will be greater for a Hankel beam with a higher power density. For example, from a comparison of Figs. 5.2 *c* and 5.2 *f* it can be seen that the maximum intensity for the H2-beam is more than two times greater than the maximum intensity of the H1-beam. Therefore, the density of the OAM on the bright ring of the H2-beam will be as many times greater than the density of the OAM on the bright ring of the H1-beam.

The process of taking the derivatives with respect to *z* from the amplitude of the Hankel beams can be continued to obtain higher type Hankel beams. But for further consideration it was enough to get only three complex amplitudes (5.5), (5.9) and (5.19), through which the components of the electromagnetic field of the Hankel vector beam [230] are expressed.

In this section the complex amplitudes of nonparaxial scalar strongly divergent Hankel laser wave beams of the 1st (5.5), 2nd (5.9), and 3rd (5.19) types are obtained in explicit form. All three types of beams have the same whole topological charge, but a different dependence of the amplitude on the radial coordinate. All three types of Hankel beams are obtained from each other by differentiating their amplitudes along the longitudinal coordinate. The intensity of all three types of Hankel beams has a radial symmetry (in the form of a 'doughnut') and has no side lobes. With an increase in the number (from 1 to 3) of the Hankel beam type, its divergence decreases. These beams are not free-space modes, since the propagation changes the intensity distribution in the beam section, but the intensity structure (the annular intensity view without side lobes) is preserved.

5.2. Hankel beams with linear polarization

The laser beams with an orbital angular momentum (OAM) are intensively studied in optics. They are used for wireless communications with an increased information channel density [234], for optical capture and rotation of microparticles [235]. Such beams have 'immunity' to turbulence and can be used for atmospheric sensing [236]. In micromanipulation tasks it is important that the focal spot does not have side lobes [232]. Such a requirement is satisfied, for example, by the recently discovered Hankel laser beams [233]. These are nonparaxial scalar vortex laser beams, which, like ordinary Gaussian beams, do not have side lobes. Also in micromanipulation problems it is required to ensure sharp focussing of the laser beam. With a sharp focussing of light, when the rays converge into focus at large angles, polarization begins to play an important role in the formation of the focus. For an adequate description of the OAM in the field of focus, it is required to consider all six projections of the electromagnetic field intensity vectors. In optics, a small number of vector light fields are known, which can be described analytically explicitly. Closed analytical expressions for the OAM have so far been obtained only for the paraxial Laguerre–Gauss beams [9] and for the Bessel vector beams [131]. In [230], expressions for the projections of the electric (E-vector) and magnetic (H-vector) intensity vectors of electromagnetic wave fields were obtained. All projections of the E- and H-vectors are expressed through three integrals, which are expansions in the angular spectrum of the plane waves. On the basis of projections of the electromagnetic field it is possible to obtain the expressions for the Poynting vector and the OAM are obtained.

In this section attention is given to new vector nonparaxial vortex Hankel beams. For the Hankel vector beam with linear polarization, explicit analytical expressions are obtained for all projections of the E- and H-vectors of the electromagnetic wave intensity. On the basis of the formulas obtained, one can analytically obtain an expression for the density vector of the orbital angular momentum of the Hankel vector beams.

In [230] it is shown that the vector of the angular momentum (AM), equal to

$$\mathbf{j} = \varepsilon\mu[\mathbf{r} \times \mathbf{S}], \tag{5.20}$$

where ε is the dielectric constant of the medium, μ is the magnetic permeability of the medium, $\mathbf{S}$ is the Umov–Poynting vector

$$\mathbf{S} = \frac{1}{2}\left([\mathbf{E} \times \mathbf{H}^*] + [\mathbf{E}^* \times \mathbf{H}]\right), \tag{5.21}$$

where $\mathbf{E}$ and $\mathbf{H}$ are the vectors of the electric and magnetic field strengths of an electromagnetic wave can be calculated for a wave with linear polarization directed along the x axis using the following expressions for all projections of the field strength vectors:

$$E_x = I_1, \quad E_y = 0, \quad E_z = -\frac{1}{k}\frac{\partial I_2}{\partial x},$$

$$H_x = \frac{i}{\mu k^2}\frac{\partial^2 I_2}{\partial y \partial x},$$

$$H_y = \frac{i}{\mu}I_3 + \frac{i}{\mu k^2}\frac{\partial^2 I_2}{\partial x^2}, \tag{5.22}$$

$$H_z = \frac{i}{\mu k}\frac{\partial I_1}{\partial y}.$$

The following integrals are used in (5.22):

$$I_1 = \int_0^\infty \int_0^{2\pi} A(\rho,\theta)\exp\left[ikr\rho\cos(\varphi-\theta) - kz\sqrt{\rho^2-1}\right]\rho\,d\rho\,d\theta, \tag{5.23}$$

$$I_2 = \int_0^\infty \int_0^{2\pi} \frac{A(\rho,\theta)}{\sqrt{\rho^2-1}} \exp\left[ikr\rho\cos(\varphi-\theta) - kz\sqrt{\rho^2-1}\right]\rho \, d\rho \, d\theta, \qquad (5.24)$$

$$I_3 = \int_0^\infty \int_0^{2\pi} A(\rho,\theta)\sqrt{\rho^2-1} \exp\left[ikr\rho\cos(\varphi-\theta) - kz\sqrt{\rho^2-1}\right]\rho \, d\rho \, d\theta, \qquad (5.25)$$

where $A(\rho,\theta)$ is the complex amplitude of the spectrum of plane waves, (x,y,z) are the Cartesian coordinates, z is the coordinate along the optical axis, (r,φ,z) are the cylindrical coordinates and (ρ,θ) is the dimensionless polar coordinates of the spectrum of plane waves (that is, ρ is the ratio of the projection of the wave vector k on the transverse plane to the wave number k).

The double integrals (5.23)–(5.25) are reduced to single integrals, if one can explicitly single out the 'vortex' component of the amplitude of the spectrum of plane waves:

$$A(\rho,\theta) = A(\rho)\exp(in\theta), \qquad (5.26)$$

where n is an integer. Taking into account (5.26), expressions (5.23)–(5.25) will look like:

$$I_1 = 2\pi i^n e^{in\phi} \int_0^\infty A(\rho)\exp\left(-kz\sqrt{\rho^2-1}\right) J_n(kr\rho)\rho \, d\rho, \qquad (5.27)$$

$$I_2 = 2\pi i^n e^{in\phi} \int_0^\infty \frac{A(\rho)}{\sqrt{\rho^2-1}}\exp\left(-kz\sqrt{\rho^2-1}\right) J_n(kr\rho)\rho \, d\rho, \qquad (5.28)$$

$$I_3 = 2\pi i^n e^{in\phi} \int_0^\infty A(\rho)\sqrt{\rho^2-1}\exp\left(-kz\sqrt{\rho^2-1}\right) J_n(kr\rho)\rho \, d\rho, \qquad (5.29)$$

where $J_n(x)$ is the Bessel function of the first kind of the n-th order. Note that the expressions (5.27)–(5.29) are related to each other using the derivatives:

$$I_3 = -\frac{1}{k}\frac{\partial I_1}{\partial z} = \frac{1}{k^2}\frac{\partial^2 I_2}{\partial z^2} \quad , \quad I_1 = -\frac{1}{k}\frac{\partial I_2}{\partial z}. \qquad (5.30)$$

Let $A(\rho) = \rho^n$. In this case, the main contribution to the field is made by evanescent waves. The power of such a spectrum of plane waves is unlimited, and also the power of the light field having such a spectrum is unlimited. If the spectrum of plane waves is given by such a power function, the integrals (5.27)–(5.29) can be calculated using the Hankel functions:

$$I_1 = 2\pi i^n e^{in\phi} \int_0^\infty \rho^n \exp\left(-kz\sqrt{\rho^2-1}\right) J_n(kr\rho)\rho d\rho =$$
$$= i^{n-1}\pi\lambda^{1/2} z r^n e^{in\phi} \Psi_{n+3/2}(R),$$

(5.31)

$$I_2 = 2\pi i^n e^{in\phi} \int_0^\infty \frac{\rho^n}{\sqrt{\rho^2-1}} \exp\left(-kz\sqrt{\rho^2-1}\right) J_n(kr\rho)\rho d\rho =$$
$$= i^{n-1}\pi\lambda^{1/2} r^n e^{in\phi} \Psi_{n+1/2}(R),$$

(5.32)

$$I_3 = 2\pi i^n e^{in\phi} \int_0^\infty \rho^n \sqrt{\rho^2-1} \exp\left(-kz\sqrt{\rho^2-1}\right) J_n(kr\rho)\rho d\rho =$$
$$= i^{n-1}\pi\lambda^{1/2} r^n e^{in\phi} \left[z^2 \Psi_{n+5/2}(R) - k^{-1}\Psi_{n+3/2}(R)\right],$$

(5.33)

$$\Psi_{n+v}(R) = \frac{H_{n+v}^{(2)}(kR)}{R^{n+v}}, \quad R = \sqrt{r^2+z^2},$$

(5.34)

$$\frac{\partial \Psi_{n+v}(R)}{\partial x} = -\Psi_{n+v+1}(R)(kr\cos\phi),$$
$$\frac{\partial \Psi_{n+v}(R)}{\partial y} = -\Psi_{n+v+1}(R)(kr\sin\phi),$$

(5.35)

where $H_v^{(2)}(x) = J_v(x) - iY_v(x)$ is the Hankel function, $Y_v(x)$ is the Neumann function.

It can be seen from (5.31)–(5.33) that the integrals I_1, I_2, I_3 are equal, respectively, to the Hankel beams of the 2nd, 1st, and 3rd types [233]. The auxiliary equations (5.35) are used to obtain the following formulas.

Using (5.23) and (5.31)–(5.34), we find all the projections of the Hankel vector light field with linear polarization ($E_y(r,\varphi,z) = 0$):

$$E_x(r,\phi,z) = i^{n+1}\pi\lambda^{1/2} z r^n e^{in\phi} \Psi_{n+3/2}(R),$$

(5.36)

$$E_z(r,\phi,z) = \frac{i^{n+1}\lambda^{3/2}r^{n-1}e^{in\phi}}{2} \times$$
$$\times \left[ne^{-i\phi}\Psi_{n+1/2}(R) - kr^2\cos\phi\,\Psi_{n+3/2}(R)\right], \tag{5.37}$$

$$H_x(r,\phi,z) = \frac{i^n\lambda^{5/2}r^{n-2}e^{in\phi}}{4\pi\mu}\left[in(n-1)e^{-i2\phi}\Psi_{n+1/2}(R) - \right.$$
$$\left. -inkr^2e^{-i2\phi}\,\Psi_{n+3/2}(R) + k^2r^4\sin\phi\cos\phi\,\Psi_{n+5/2}(R)\right], \tag{5.38}$$

$$H_y(r,\phi,z) = \frac{i^n\lambda^{5/2}r^{n-2}e^{in\phi}}{4\pi\mu}\left[n(n-1)e^{-i2\phi}\Psi_{n+1/2}(R) - \right.$$
$$\left. -2kr^2(1 + ne^{-i\phi}\cos\phi)\Psi_{n+3/2}(R) + k^2r^2(z^2 + r^2\cos^2\phi)\Psi_{n+5/2}(R)\right], \tag{5.39}$$

$$H_z(r,\phi,z) = \frac{i^n\lambda^{3/2}zr^{n-1}e^{in\phi}}{2\mu}\left[ine^{-i\phi}\Psi_{n+3/2}(R) - kr^2\sin\phi\,\Psi_{n+5/2}(R)\right], \tag{5.40}$$

From (5.20) the projection on the z axis of the AM will be:

$$j_z = \frac{\varepsilon\mu^2}{2}\left[x\,\mathrm{Re}(E_zH_x^* - H_z^*E_x) + y\,\mathrm{Re}(E_zH_y^*)\right]. \tag{5.41}$$

From (5.41), it is clear that the expressions (5.36)–(5.40) allow in a closed form (albeit cumbersome) to obtain the density of the projection of the AM onto the optical axis for the nonparaxial Hankel vector beam.

From (5.36) it can be seen that the transverse projection of the E-field has a radially symmetric intensity distribution with zero intensity on the optical axis (with $r = 0$), except for the initial plane ($z = 0$) and except $n = 0$, when the Hankel beam coincides with a spherical wave. From (5.37) it can be seen that the longitudinal projection of the E-field will not have radial symmetry, but will have symmetry with respect to both Cartesian coordinates, and will be elongated along the polarization direction (along the x axis). Also, the longitudinal projection of the E-field on the axis will always be zero, except in the case of $n = 1$. When $n = 1$ at the centre of the intensity distribution is a saddle point: a minimum in the x coordinate and a maximum in the y coordinate. This means that the total intensity of the Hankel vector beam with linear polarization $I = |E_x|^2 + |E_z|^2$ will not have a radial symmetry about the optical axis.

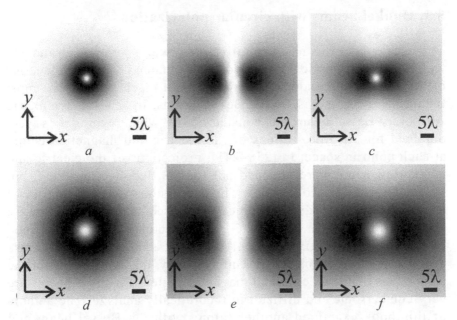

Fig. 5.3. The intensity of the (negative) of the lateral (a, d), the longitudinal (b, e) and total (c, f) of the intensity of the Hankel vector beam (wavelength $\lambda = 532$ nm, topological charge $n = 1$) in planes $z = 5\lambda$ (a, b, c) and $z = 10\lambda$ (d, e, f).

Figure 5.3 shows the distributions of the transverse and longitudinal components of the intensity, as well as the total intensity (i.e., the values $I_x = |E_x|^2$, $I_z = |E_z|^2$, $I = I_x + I_z$, respectively) of the Hankel vector beam with linear polarization along the x axis, with topological charge $n = 1$ in planes $z = 5\lambda$ and $z = 10\lambda$.

The complex amplitudes of the Hankel scalar beams obtained in [233] were used to obtain the Hankel vector beams. The amplitudes of all projections of the electric and magnetic field strength vectors (5.36)–(5.40) for the Hankel vector nonparaxial vortex beam with linear polarization were considered in explicit form. Thus, the expressions for projections of the Umov–Poynting vector (UP) (2) and AM (1) are obtained in a closed analytical form. Until now, it was possible to write analytically the projections of these UP and AM vectors only for the nonparaxial vector Bessel beam [131] and paraxial Laguerre–Gauss beams [9]. Although the intensity of the transverse electric component E_x (5.36) has a radial symmetry, the intensity of the other projections of the light field E_z (5.37), H_x (5.38), H_y (5.39) and H_z (5.40) does not have radial symmetry for the linearly polarized Hankel beam.

5.3. Hankel beams with circular polarization

The vector laser beams that satisfy Maxwell's equations and for which explicit analytical expressions are obtained for the projections of the electric and magnetic vectors are not much known. The most famous of these beams are the Bessel beams. In [237], a symmetric zero-order Bessel beam was considered, for which expressions were obtained for all 6 projections of the electric and magnetic vectors at each point in space. In [238–241], projections of the electric and magnetic vectors were obtained not for the Bessel beam itself, but for its TE and TM mode states. The expressions for a vector Bessel beam of arbitrary order with linear and circular polarizations were obtained in [242]. In [242], the Bessel vector beam was obtained by the Hertz potential method. In [243], spherical quasi-Gauss vector beams were considered. Unfortunately, the expressions for the projections of the electric and magnetic vectors for these beams are very cumbersome and difficult to theoretically analyze. The authors of this book examined another (along with the Bessel beam and the spherical quasi-Gauss beam) vector beam, which is amenable to a complete analytical description. Nonparaxial scalar [233] and vector [244, 245] Hankel beams with linear polarization were also investigated. These beams were obtained by plane wave expansion [230]. Vector cylindrical beams with radial or azimuthal polarizations are also known [246]. However, no expressions were obtained for the projections of the electric and magnetic vectors for such beams. Cylindrical beams are analyzed using Richards–Wolf formulas and are used for sharp focussing of light [247].

Explicit formulas for the projections of the electric and magnetic vectors of the laser beam make it possible to obtain analytical expressions for the Pointing vector and the angular momentum of the light field [9, 131, 238, 245].

Explicit expressions were obtained for the projections of the electric and magnetic vectors of the Hankel beams with right and left circular polarization. From the obtained expressions, we can conclude that for a Hankel beam with a topological charge n, the amplitude of the longitudinal components of the electric and magnetic fields has a topological charge of $n + 1$ for the right and $n-1$ for the left circular polarization.

5.3.1. Projections of electromagnetic field vectors for right and left circular polarization

Similarly, the method of calculating, described in [233, 244], can be used to find all the projections of the electric $\mathbf{E} = (E_x, E_y, E_z)$ and magnetic $\mathbf{H} = (H_x, H_y, H_z)$ vectors of a Hankel beam with right $E_x + iE_y$ and $E_x - iE_y$ left circular polarization. We will denote the amplitudes of the projections of the electric and magnetic vectors by superscripts $+$ and $-$ to denote the right and left circular polarization. Then for the Hankel beam with the right polarization we get:

$$E_{ny}^{+}(r,\varphi,z) = iE_{nx}(r,\varphi,z) = i^n \pi \lambda^{1/2} z \left(re^{i\varphi} \right)^n \psi_{n+3/2}(R), \tag{5.42}$$

$$\psi_{n+\nu}(R) = \frac{H_{n+\nu}^{(1)}(kR)}{R^{n+\nu}}, \quad R = \left(z^2 + r^2 \right)^{1/2}, \tag{5.43}$$

$$E_{nz}^{+}(r,\varphi,z) = \frac{-i}{k} \left(i\frac{\partial}{\partial x} - \frac{\partial}{\partial y} \right) \left(i^{n-1} \pi \lambda^{1/2} \left(re^{i\varphi} \right)^n \psi_{n+1/2}(R) \right) = \tag{5.44}$$

$$= i^{n+1} \pi \lambda^{1/2} \left(re^{i\varphi} \right)^{n+1} \psi_{n+3/2}(R),$$

$$H_{nx}^{+}(r,\varphi,z) = \frac{i}{k\mu} \left(\frac{\partial E_z^{+}}{\partial y} - i\frac{\partial E_x}{\partial z} \right) = \tag{5.45}$$

$$= \frac{i^{n-1} \lambda^{3/2}}{2\mu} \left(re^{i\varphi} \right)^n \left\{ (n+2) \psi_{n+3/2}(R) - k\left[z^2 - ir^2 e^{i\varphi} \sin\varphi \right] \psi_{n+5/2}(R) \right\},$$

$$H_{ny}^{+}(r,\varphi,z) = \frac{i}{k\mu} \left(-\frac{\partial E_z^{+}}{\partial x} + \frac{\partial E_x}{\partial z} \right) = \tag{5.46}$$

$$= \frac{i^n \lambda^{3/2}}{2\mu} \left(re^{i\varphi} \right)^n \left[(n+2) \psi_{n+3/2}(R) - k\left(z^2 + r^2 e^{i\varphi} \cos\varphi \right) \psi_{n+5/2}(R) \right],$$

$$H_{nz}^{+}(r,\varphi,z) = \frac{i}{k\mu} \left(i\frac{\partial}{\partial x} - \frac{\partial}{\partial y} \right) \left(i^{n-1} \pi \lambda^{1/2} z \left(re^{i\varphi} \right)^n \psi_{n+3/2}(R) \right) = \tag{5.47}$$

$$= \frac{i^{n-1} \pi \lambda^{1/2} z}{\mu} \left(re^{i\varphi} \right)^{n+1} \psi_{n+5/2}(R),$$

For the Hankel beam with left polarization we get:

$$E_{ny}^{-}(r,\varphi,z) = -iE_{nx}(r,\varphi,z) = -i^n \pi \lambda^{1/2} z \left(re^{i\varphi} \right)^n \psi_{n+3/2}(R), \tag{5.48}$$

$$E_{nz}^{-}(r,\varphi,z)=\frac{-i}{k}\left(i\frac{\partial}{\partial x}+\frac{\partial}{\partial y}\right)\left(i^{n-1}\pi\lambda^{1/2}\left(re^{i\varphi}\right)^{n}\psi_{n+1/2}(R)\right)=$$

$$=i^{n-1}\lambda^{3/2}\left(re^{i\varphi}\right)^{n-1}\left[n\psi_{n+1/2}(R)-\frac{kr^{2}}{2}\psi_{n+3/2}(R)\right],$$

(5.49)

$$H_{nx}^{-}(r,\varphi,z)=\frac{i}{k\mu}\left(\frac{\partial E_{z}^{-}}{\partial y}+i\frac{\partial E_{x}}{\partial z}\right)=$$

$$=\frac{i^{n+1}\lambda^{3/2}}{2\mu}\left(re^{i\varphi}\right)^{n}\left\{\frac{2n(n-1)e^{-2i\varphi}}{kr^{2}}\psi_{n+1/2}(R)-\right.$$

(5.50)

$$\left.-\left(2ne^{-2i\varphi}-n-2\right)\psi_{n+3/2}(R)-k\left(z^{2}+ir^{2}e^{-i\varphi}\sin\varphi\right)\psi_{n+5/2}(R)\right\},$$

$$H_{ny}^{-}(r,\varphi,z)=\frac{i}{k\mu}\left(-\frac{\partial E_{z}^{-}}{\partial x}+\frac{\partial E_{x}}{\partial z}\right)=$$

$$=\frac{i^{n}\lambda^{3/2}}{2\mu}\left(re^{i\varphi}\right)^{n}\left\{-\frac{2n(n-1)e^{-2i\varphi}}{kr^{2}}\psi_{n+1/2}(R)+\right.$$

(5.51)

$$\left.+\left(2ne^{-2i\varphi}+n+2\right)\psi_{n+3/2}(R)-k\left(z^{2}+r^{2}e^{-i\varphi}\cos\varphi\right)\psi_{n+5/2}(R)\right\},$$

$$H_{nz}^{-}(r,\varphi,z)=\frac{-i}{k\mu}\left(i\frac{\partial}{\partial x}+\frac{\partial}{\partial y}\right)\left[i^{n-1}\pi\lambda^{1/2}z\left(re^{i\varphi}\right)^{n}\psi_{n+3/2}(R)\right]=$$

$$=\frac{i^{n-1}\lambda^{3/2}z}{\mu}\left(re^{i\varphi}\right)^{n-1}\left[n\psi_{n+3/2}(R)-\frac{kr^{2}}{2}\psi_{n+5/2}(R)\right].$$

(5.52)

Comparing (5.42)–(5.47) and (5.48)–(5.52) shows that the right circular polarization with a positive topological charge increases this charge by one for the amplitudes of the longitudinal projections of the electric and magnetic vectors, E_{nz}^{+} and H_{nz}^{+}, conversely, the left circular polarization decreases by unit topological charge of the optical vortex n near the amplitudes of the longitudinal projections of the electric E_{nz}^{-} and H_{nz}^{-} magnetic vectors. For the remaining projections, the topological charge does not change and remains equal to n. This leads to the fact that in the absence of an optical vortex ($n=0$), the longitudinal components of the Hankel beam with left and right polarizations are optical vortices with topological charges $n=\pm1$:

$$E_{0z}^{\pm}\left(r,\varphi,z\right)=i\pi\lambda^{1/2}re^{\pm i\varphi}\psi_{3/2}\left(R\right), \tag{5.53}$$

$$H_{0z}^{\pm}\left(r,\varphi,z\right)=\mp\frac{i\pi\lambda^{1/2}z}{\mu}re^{\pm i\varphi}\psi_{5/2}\left(R\right) \tag{5.54}$$

and on the optical axis $(r = 0)$ only the transverse components of the electromagnetic field are different from 0 $(E_{0z}^{\pm}(r=0,\varphi,z)=H_{0z}^{\pm}(r=0,\varphi,z)=0)$:

$$\begin{aligned}E_{0y}^{+}\left(r=0,\varphi,z\right)=iE_{0x}\left(r=0,\varphi,z\right)=\\-E_{0y}^{-}\left(r=0,\varphi,z\right)=\pi\lambda^{1/2}z\psi_{3/2}\left(z\right),\end{aligned} \tag{5.55}$$

$$H_{0x}^{+}\left(r=0,\varphi,z\right)=-H_{0x}^{-}\left(r=0,\varphi,z\right)=\frac{-i\lambda^{3/2}}{2\mu}\left[2\psi_{3/2}\left(z\right)-kz^{2}\psi_{5/2}\left(z\right)\right], \tag{5.56}$$

$$H_{0y}^{+}\left(r=0,\varphi,z\right)=H_{0y}^{-}\left(r=0,\varphi,z\right)=\frac{\lambda^{3/2}}{2\mu}\left[2\psi_{3/2}\left(z\right)-kz^{2}\psi_{5/2}\left(z\right)\right], \tag{5.57}$$

From (5.56)–(5.57) it can be seen that at some distances z, the transverse magnetic components can be zero. With $n > 1$ on the optical axis $(r = 0)$ all the projections of the Hankel beam field are zero. When $n = 1$ on the optical axis $(r = 0)$, only the longitudinal components of the Hankel beam with left polarization are different from zero

$$E_{1z}^{-}\left(r=0,\varphi,z\right)=\lambda^{3/2}\psi_{3/2}\left(z\right), \tag{5.58}$$

$$H_{1z}^{-}\left(r=0,\varphi,z\right)=\frac{\lambda^{3/2}z}{\mu}\psi_{5/2}\left(R\right). \tag{5.59}$$

and when $n = -1$ on the optical axis $(r = 0)$, only the longitudinal components of the Hankel beam with the right polarization are different from zero

$$E_{-1z}^{+}\left(r=0,\varphi,z\right)=\pi\lambda^{1/2}\psi_{1/2}\left(z\right), \tag{5.60}$$

$$H_{-1z}^{+}\left(r=0,\varphi,z\right)=\frac{-\pi\lambda^{1/2}z}{\mu}\psi_{3/2}\left(z\right), \tag{5.61}$$

This leads to the fact that when $n = 1$ and left circular polarization at small distances z, a light spot forms instead of a light ring, which is then converted into a ring. This is seen in Fig. 5.4, which shows the intensity distribution of the Hankel beam with a single topological charge ($n = 1$) with a linear ($E_y = 0$) (Fig. 5.4 *a*, *d*), right circular (Fig. 5.4 *b*, *e*) and left circular (Fig. 5.4 *c*, *f*) polarization at distances $z = \lambda/4$ (Fig. 5.4 *a–c*) and $z = \lambda/2$ (Fig. 5.4 *d–f*). In Fig.

5.4 a–c shows the region $-\lambda/2 \le x, y \le \lambda/2$, and Fig. 5.4 d–e shows the region $-3\lambda/2 \le x, y \le 3/2$.

It is seen from Fig. 5.4 that if for the linear polarization the Hankel beam does not have circular symmetry (and rotates in the near zone), then the Hankel beams with circular polarization have circular symmetry and look like a circle of light or a ring. The latter can also be seen from the expression for the intensity for the right circular polarization, which does not depend on the azimuth angle:

$$I_n^+(r,z) = \pi^2 \lambda r^{2n}\left(2z^2 + r^2\right)\left|\psi_{n+3/2}(R)\right|^2. \tag{5.62}$$

The expression for the intensity of the Hankel beam with the left circular polarization is more cumbersome, but also does not depend on the azimuth angle:

$$I_n^-(r,z) = 2\pi^2 \lambda z^2 r^{2n}\left|\psi_{n+3/2}(R)\right|^2 + \lambda^3 r^{2n-2} \times$$
$$\times\left[n^2\left|\psi_{n+1/2}(R)\right|^2 + \frac{k^2 r^4}{2}\left|\psi_{n+3/2}(R)\right|^2 - 2nkr^2 \operatorname{Re}\left(\psi_{n+1/2}(R)\psi^*_{n+3/2}(R)\right) \right], \tag{5.63}$$

where Re $(...)$ is the real part of the number. From (5.63) it can be

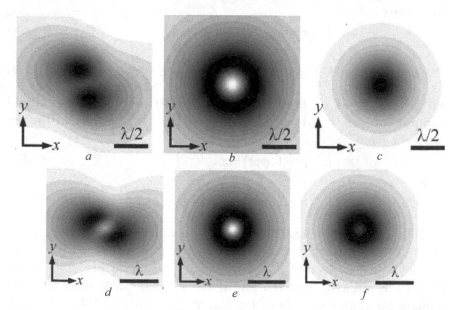

Fig. 5.4. Intensity distribution $I = \left|E_x\right|^2 + \left|E_y\right|^2 + \left|E_z\right|^2$ (negative) of the Hankel beam ($n = 1$) with linear (a, d), right circular (b, e) and left circular (c, f) polarization at distances $z = \lambda/4$ (a–c) and $z = \lambda/2$ (d–f).

seen that when $n = 1$ on the optical axis ($r = 0$), the intensity will decrease with increasing z:

$$I_1^-(r=0,z) = \lambda^3 \left|\psi_{3/2}(z)\right|^2 = \frac{\lambda^4}{\pi^2 z^4}\left(1 + \frac{1}{(kz)^2}\right). \qquad (5.64)$$

5.3.2. Hankel vortex vector beams with circular polarization in the far zone

When $z \gg \lambda$ the formulas are simplified. Using the asymptotics of the Hankel function ($x \gg 1$)

$$H_\nu^{(1)}(x) \approx \sqrt{\frac{2}{\pi x}}(-i)^{\nu+1/2}e^{ix}, \qquad (5.65)$$

instead of (5.43) we get

$$\psi_{n+\nu}(R) = \frac{H_{n+\nu}^{(1)}(kR)}{R^{n+\nu}} \approx \frac{\lambda^{1/2}(-i)^{n+\nu+1/2}e^{ikR}}{\pi R^{n+\nu+1/2}}. \qquad (5.66)$$

Substituting (5.66) into (5.42)–(5.47), for the Hankel beam with the right polarization we get:

$$E_{ny}^+(r,\varphi,z) = iE_{nx}(r,\varphi,z) = -\lambda \frac{r^n z}{R^{n+2}}e^{in\varphi+ikR}, \qquad (5.67)$$

$$E_{nz}^+(r,\varphi,z) = -i\lambda \frac{r^{n+1}}{R^{n+2}}e^{i(n+1)\varphi+ikR}, \qquad (5.68)$$

$$H_{nx}^+(r,\varphi,z) = \frac{\lambda}{\mu}\left(i\frac{n+2}{kR^{n+2}} + \frac{ir^2 e^{i\varphi}\sin\varphi - z^2}{R^{n+3}}\right)r^n e^{in\varphi+ikR}, \qquad (5.69)$$

$$H_{ny}^+(r,\varphi,z) = i\frac{\lambda}{\mu}\left(i\frac{n+2}{kR^{n+2}} - \frac{r^2 e^{i\varphi}\cos\varphi + z^2}{R^{n+3}}\right)r^n e^{in\varphi+ikR}, \qquad (5.70)$$

$$H_{nz}^+(r,\varphi,z) = \frac{\lambda}{\mu}\frac{r^{n+1}z}{R^{n+3}}e^{i(n+1)\varphi+ikR}. \qquad (5.71)$$

Similarly, instead of (5.48)–(5.52) for the Hankel beam with left polarization, we obtain in the far zone:

$$E_{nz}^-(r,\varphi,z) = -i\lambda \frac{r^{n+1}}{R^{n+2}}\left(1 - \frac{2inR}{kr^2}\right)e^{i(n-1)\varphi+ikR}, \qquad (5.73)$$

$$H_{nx}^-(r,\varphi,z) = \frac{\lambda^2}{2\pi\mu} \frac{r^n}{R^{n+3}} \left\{ \frac{2n(n-1)e^{-2i\varphi}}{kr^2}R^2 + \right.$$

$$\left. +i\left(2ne^{-2i\varphi} - n - 2\right)R + k\left(z^2 + ir^2e^{-i\varphi}\sin\varphi\right)\right\}e^{in\varphi+ikR},$$

(5.74)

$$H_{ny}^-(r,\varphi,z) = \frac{\lambda^2}{2\pi\mu} \frac{r^n}{R^{n+3}} \left\{ i\frac{2n(n-1)e^{-2i\varphi}}{kr^2}R^2 + \right.$$

$$\left. +\left(2ne^{-2i\varphi} + n + 2\right)R - ik\left(z^2 + r^2e^{-i\varphi}\cos\varphi\right)\right\}e^{in\varphi+ikR},$$

(5.75)

$$H_{nz}^-(r,\varphi,z) = \frac{\lambda}{\mu} \frac{r^{n+1}z}{R^{n+3}} \left(\frac{2inR}{kr^2} - 1 \right) e^{i(n-1)\varphi+ikR}.$$

(5.76)

The intensity distribution in the far zone for the right and left polarizations instead of (5.62) and (5.63) will be equal to:

$$I_n^+(r,\varphi,z) = \left|E_{nx}^+(r,\varphi,z)\right|^2 + \left|E_{ny}^+(r,\varphi,z)\right|^2 + \left|E_{nz}^+(r,\varphi,z)\right|^2 =$$

$$= \lambda^2 \frac{r^{2n}}{R^{2n+4}}\left(r^2 + 2z^2\right),$$

(5.77)

$$I_n^-(r,\varphi,z) = \left|E_{nx}^-(r,\varphi,z)\right|^2 + \left|E_{ny}^-(r,\varphi,z)\right|^2 + \left|E_{nz}^-(r,\varphi,z)\right|^2 =$$

$$= \lambda^2 \frac{r^{2n}}{R^{2n+4}}\left(r^2 + 2z^2 + \frac{4n^2R^2}{k^2r^2}\right).$$

(5.78)

The maximum intensity for the right circular polarization lies on the light ring with a radius equal to:

$$r_{\max}^+ = z\sqrt{\frac{n-3+\sqrt{n^2+2n+9}}{2}}.$$

(5.79)

This radius is zero (that is, the intensity maximum on the optical axis) only for $n = 0$. For large topological charges ($n \gg 1$), the radius of the light ring on which the intensity maximum is located increases with increasing z as $r_{\max}^+ = z\sqrt{n}$. In the case of left circular polarization, the equation for the intensity maximum is more complicated:

$$(n-3)\frac{r^4}{z^4} + 2n\frac{r^2}{z^2} - \frac{r^6}{z^6} + \frac{4n^2(n-3)}{(kz)^2}\frac{r^2}{z^2} + \frac{4n^2(n-1)}{(kz)^2} - \frac{8n^2}{(kz)^2}\frac{r^4}{z^4} = 0, \quad (5.80)$$

but since it is assumed that $z \gg \lambda$, the denominators in the last three terms are large and therefore only the first three terms can be left. In this case, from (5.80) we obtain the following expression for the intensity maximum, which coincides with the expression for the right circular polarization:

$$r_{max}^- = z\sqrt{\frac{n-3+\sqrt{n^2+2n+9}}{2}}. \tag{5.81}$$

The Umov–Poynting vector $\mathbf{S} = \mathbf{E}^* \times \mathbf{H}$ can be calculated for Hankel beams with circular polarization using expressions for the projections of the electric and magnetic vectors (5.42)–(5.52). But these expressions will be cumbersome, and their analysis will be difficult. The expressions for the projections of the Umov–Poynting vector for the Hankel beam with circular polarization in the far zone look less cumbersome. We present expressions only for the transverse projections of the vector $\mathbf{S}$, which are involved in determining the axial projection of the angular momentum vector. Then for the right circular polarization:

$$S_{nx}^+(r,\varphi,z) = E_{ny}^{+*}(r,\varphi,z)H_{nz}^+(r,\varphi,z) - E_{nz}^{+*}(r,\varphi,z)H_{ny}^+(r,\varphi,z) =$$

$$= \left[-\lambda \frac{r^n z}{R^{n+2}} e^{-in\varphi-ikR} \right]\left[\frac{\lambda}{\mu}\frac{r^{n+1}z}{R^{n+3}} e^{i(n+1)\varphi+ikR} \right] -$$

$$- \left[i\lambda \frac{r^{n+1}}{R^{n+2}} e^{-i(n+1)\varphi-ikR} \right]\left[i\frac{\lambda}{\mu}\left(i\frac{n+2}{kR^{n+2}} - \frac{r^2 e^{i\varphi}\cos\varphi+z^2}{R^{n+3}} \right) r^n e^{in\varphi+ikR} \right] = \tag{5.82}$$

$$= \frac{\lambda^2}{\mu}\frac{r^{2n+1}}{R^{2n+5}}\left[i\frac{n+2}{k}Re^{-i\varphi} - \left(r^2+2z^2 \right)\cos\varphi \right],$$

$$S_{ny}^+(r,\varphi,z) = E_{nz}^{+*}(r,\varphi,z)H_{nx}^+(r,\varphi,z) - E_{nx}^*(r,\varphi,z)H_{nz}^+(r,\varphi,z) =$$

$$= \left[i\lambda \frac{r^{n+1}}{R^{n+2}} e^{-i(n+1)\varphi-ikR} \right]\left[\frac{\lambda}{\mu}\left(i\frac{n+2}{kR^{n+2}} + \frac{ir^2 e^{i\varphi}\sin\varphi-z^2}{R^{n+3}} \right) r^n e^{in\varphi+ikR} \right] +$$

$$+ \left[i\lambda \frac{r^n z}{R^{n+2}} e^{-in\varphi-ikR} \right]\left[\frac{\lambda}{\mu}\frac{r^{n+1}z}{R^{n+3}} e^{i(n+1)\varphi+ikR} \right] = \tag{5.83}$$

$$= \frac{\lambda^2}{\mu}\frac{r^{2n+1}}{R^{2n+5}}\left[-\frac{n+2}{k}Re^{-i\varphi} - \left(r^2+2z^2 \right)\sin\varphi \right],$$

and for left circular polarization:

$$S_{nx}^{-}(r,\varphi,z) = E_{ny}^{-*}(r,\varphi,z)H_{nz}^{-}(r,\varphi,z) - E_{nz}^{-*}(r,\varphi,z)H_{ny}^{-}(r,\varphi,z) =$$

$$= \frac{\lambda^2}{\mu}\frac{r^{2n+1}}{R^{2n+5}}\left\{2z^2\left(\frac{2nR}{kr^2}\sin\varphi - \cos\varphi\right) + \left(1+\frac{2inR}{kr^2}\right)\times \right.$$
$$\left. \times\left[\frac{2n(n-1)e^{-i\varphi}}{r^2}\frac{R^2}{k^2} + i\left(2ne^{-2i\varphi}+n+2\right)\frac{R}{k}e^{i\varphi} - r^2\cos\varphi\right]\right\} \quad (5.84)$$

$$S_{ny}^{-}(r,\varphi,z) = E_{nz}^{-*}(r,\varphi,z)H_{nx}^{-}(r,\varphi,z) - E_{nx}^{-*}(r,\varphi,z)H_{nz}^{-}(r,\varphi,z) =$$

$$= \frac{\lambda^2}{\mu}\frac{r^{2n+1}}{R^{2n+5}}\left\{-2z^2\left(\sin\varphi + \frac{2nR}{kr^2}\cos\varphi\right) + \left(1+\frac{2inR}{kr^2}\right)\times \right.$$
$$\left. \times\left[i\frac{2n(n-1)e^{-i\varphi}}{r^2}\frac{R^2}{k^2} - \left(2ne^{-2i\varphi}-n-2\right)\frac{R}{k}e^{i\varphi} - r^2\sin\varphi\right]\right\}. \quad (5.85)$$

The projections on the z axis of the angular momentum density vectors $\mathbf{j} = \mathrm{Re}[\mathbf{r}\times\mathbf{S}]$ for Hankel beams in the far zone with right and left circular polarizations have the form:

$$j_{nz}^{+}(r,\varphi,z) = \mathrm{Re}\left\{xS_{ny}^{+}(r,\varphi,z) - yS_{nx}^{+}(r,\varphi,z)\right\} = -\frac{\lambda^2}{\mu}\frac{n+2}{k}\frac{r^{2n+2}}{R^{2n+4}}. \quad (5.86)$$

$$j_{nz}^{-}(r,\varphi,z) = \mathrm{Re}\left\{xS_{ny}^{-}(r,\varphi,z) - yS_{nx}^{-}(r,\varphi,z)\right\} =$$
$$= -\frac{\lambda^2}{\mu}\frac{r^{2n+1}}{R^{2n+5}}\left[4n^2(n-1)\frac{R^3}{k^3r^3} + (n-2)\frac{R}{k}r + 4nz^2\frac{R}{kr}\right]. \quad (5.87)$$

It can be shown that for $r = 0$, the density of the OAM (5.86), (5.87) does not tend to infinity, but is zero regardless of polarization. From (5.86) it follows that the full angular momentum of the Hankel beam is infinite. But if the infinite total angular momentum is divided into infinite total intensity or infinite full axial projection of the Poynting vector, then the resulting normalized angular momentum must be finite. Note that for $n = 0$, the Hankel beam has a spin angular momentum:

$$j_{0z}^{+}(r,\varphi,z) = -j_{0z}^{-}(r,\varphi,z) = -\frac{2\lambda^2}{\mu k}\frac{r^2}{R^4}. \quad (5.88)$$

In this section, explicit analytical expressions are obtained for the projections of the electric magnetic vectors of Hankel vortex beams with right and left circular polarizations. It is shown that the right circular polarization with a positive topological charge increases this charge by one for the amplitudes of the longitudinal projections of the electric and magnetic vectors, and, conversely, the left circular polarization reduces by one the topological charge of the optical vortex for the amplitudes for the longitudinal projections of the electric and magnetic vectors. The remaining projections have a topological charge of n. Explicit expressions are obtained for the intensity distribution of the electric vector for Hankel beams with circular polarization, from which it follows that the Hankel beam with circular polarization has circular symmetry (in the beam section, either the light circle or the ring). For example, the Hankel beam with linear polarization does not possess circular symmetry [245]. For the Hankel beam with circular polarization in the far diffraction zone ($R \gg \lambda$), explicit analytical expressions were obtained for the transverse projections of the Umov–Poynting vector and for the axial projection of the angular momentum vector. For a beam with right circular polarization, the angular momentum is proportional to $n+2$, for a beam with left circular polarization, the angular momentum is proportional to $n-2$, and for a beam with linear circular polarization, the angular momentum is proportional to the topological charge n [245].

Conclusion

The book discusses new accelerated and vortex laser beams, which have been introduced into optics in the last 10 years. Accelerated beams are two-dimensional Airy beams. The main feature of the Airy beams is the curvature of the trajectory of the main maximum (main lobe). It propagates along a parabolic trajectory. Therefore, the Airy beams are called accelerating or ballistic (since a freely falling body moves along a parabola with constant acceleration). Other accelerated beams that propagate not along parabolic trajectories, but, for example, along a hyperbolic one, and others, are also considered. The method of transformation of slowing beams into accelerated ones is proposed. Three-dimensional Pearcey beams and Pearcey half- beams, which have the property of sharp (accelerated) self-focussing, are considered. Vortex laser beams were previously considered by the authors in the monograph 'Vortex laser beams' (Samara, Novaya teknika, 2012). In this book, new laser beams are considered that are not included in the monograph mentioned above. These are Hermite– Gauss vortex beams, asymmetric non-diffraction Bessel and Lommel modes, paraxial asymmetric Bessel–Gauss and asymmetric Laguerre–Gauss beams, and Hankel vector beams with linear and circular polarization. Analytical expressions for the orbital angular momentum are found for all vortex beams. All the considered vortex beams (except for the Hankel beams) depend on a continuous parameter, with which the orbital angular moment of the beam changes continuously. It can be whole or fractional. Many of the considered vortex laser beams were formed experimentally using a liquid-crystal light modulator. Vortex laser beams with an orbital angular momentum are used to increase the capacity of the information channels in wireless communication systems (the number of degrees of freedom in a laser beam increases), in quantum computer science to form a pair of photons entangled in the orbital angular momentum (the state of a photon with a fractional orbital moment is entangled), for sensing the atmosphere in the presence of turbulence (vortex beams are more resistant to distortion than the usual Gaussian beams) to capture and rotate micro-objects, including biological ones (vortex beams transmit their orbital angular momentum of a particle trapped and rotating it). This area of optics and photonics (laser beams, singular optics) is a hot topic booming, and this book is only an introduction to it.

References

1. Durnin J., Miceli J.J., Eberly J.H. Diffraction-free beams, Phys. Rev. Lett. 1987. Vol. 58. P. 1–1501.
2. Siegman A.E. Lasers, University Science,1986.
3. Abramochkin E.G., Volostnikov V.G. Generalized Gaussian beams, J. Opt. A: Pure Appl. Opt. 2004. Vol. 6. P. 5157–5161.
4. Gutiérrez-Vega J.C., Iturbe-Castillo M.D., Ramirez G.A., Tepichin E., Rodriguez-Dagnino R.M., Cháavez-Cerda S., New G.H.C. Experimental demonstration of optical Mathieu beams, Opt. Commun. 2001. Vol. 195, No. 1. P. 35–40.
5. Bandres M.A., Gutiérrez-Vega J.C. Ince-Gaussian beams, Opt. Lett. 2004. Vol. 29, No. 2. P. 144–146.
6. Kotlyar V.V., Khonina S.N., Skidanov R.V., Soifer V.A. Hypergeometric modes, Opt. Lett. 2007. Vol. 32, No. 7. P. 742–744.
7. Siviloglou G.A., Christodoulides D.N. Accelerating finite energy Airy beams, Opt. Lett. 2007. Vol. 32. P. 979–981.
8. Ring J., Lindberg J., Mourka A., Mazilu M., Dholakia K., Dennis M. Auto-focusing and self-healing of Pearcey beams, Opt. Express. 2012. Vol. 20. P. 18955–18966.
9. Allen L., Beijersergen M.W., Spreeuw R.J.C., Woerdman J.P. Orbital angular momentum of light and the transformation of Laguerre-Gaussian laser modes, Phys. Rev. A. 1992. Vol. 45. P. 8185–8189.
10. Kotlyar V.V., Kovalev A.A. Votex laser beams, Samara, Novaya tekhnika, 2012.
11. Zhu K., Zhou G., Li X., Zhen X., Tang H. Propagation of Bessel-Gaussian beams with optical vortices in turbulent atmosphere, Opt. Express. 2008.Vol. 16, No. 26. P. 12315–12320.
12. Krenn M., Fickler R., Fink M., Handsteiner J., Malik M., Scheidl T., Ursin R., Zeilinger A. Communication with spatially modulated light through turbulent air across Vienna, New J. Phys. 2014. Vol. 16. P. 113028.
13. Hadzievski L., Maluckov A., Rubenchik A.M., Turitsyn S. Stable optical vortices in nonliner multicore fibers, Light: Science. Appl. 2015. Vol. 4. P. e314.
14. Foo G., Palacios D.M., Swartzlander Jr. G.A. Optical vortex coronagraph, Opt. Lett. 2005. Vol 30, No. 24. P. 3308–3310.
15. Mair A., Vaziri A., Weihs G., Zeilinger A. Entanglement of the orbital angular momentum states of photons, Nature. 2001. Vol. 412, No. 6844. P. 313–316.
16. Otsu T., Ando T., Takiguchi Y., Ohtake Y., Toyoda H., Itoh H. Direct evidence for three-dimensional off-axis trapping with single Laguerre-Gaussian beam, Sci. Rep. 2014. Vol. 4. P. 4579.
17. Kalnins E.G., Miller Jr. W. Lie theory and separation of variables, J. Math. Phys. 1974. Vol. 15. P. 1728–1737.
18. Berry M.V., Balazs N.L. Nonspreading wave packets, Am. J. Phys. 1979. Vol. 47. P. 264–267.

19. Besieris I.M., Shaarawi A.M., Ziolkowski R.W. Nondispersive accelerating wave packets, Am. J. Phys. 1994. Vol. 62, No. 6. P. 519–521.

20. Besieris I.M., Shaarawi A.M. A note on an accelerating finite energy Airy beam, Opt. Lett. 2007. Vol. 32. P. 2447–2449.

21. Siviloglou G.A., Broky J., Dogariu A., Christodoulides D.N. Observation of accelerating Airy beams, Phys. Rev. Lett. 2007. Vol. 99. P. 213901.

22. Bandres M.A., Gutiérrez-Vega J.C. Airy-Gauss beams and their transformation by paraxial optical systems, Opt. Express. 2007. Vol. 15. P. 16719–16728.

23. Siviloglou G.A., Broky J., Dogariu A., Christodoulides D.N. Ballistic dynamics of Airy beams, Opt. Lett. 2008. Vol. 33. P. 207–209.

24. Sztul H.I., Alfano R.R. The Poynting vector and angular momentum of Airy beams, Opt. Express. 2008. Vol. 16. P. 9411–9416.

25. Bandres M.A. Accelerating parabolic beams, Opt. Lett. 2008. Vol. 33. P. 1678–1680.

26. Polynkin P., Kolesik M., Moloney J.V., Siviloglou G.A., Christodoulides D.N. Curved plasma channel generation using ultraintense Airy beams, Science. 2009. Vol. 324. P. 229–232.

27. Bandres M.A. Accelerating beams, Opt. Lett. 2009. Vol. 34. P. 3791–3793.

28. Carretero L., Acebal P., Blaya S., Garcia C., Fimia A., Madrigal R., Murciano A. Nonparaxial diffraction analysis of Airy and SAiry beams, Opt. Express. 2009. Vol. 17. P. 22432–22441.

29. Dolev I., Ellenbogen T., Arie A. Switching the acceleration direction of Airy beams by a nonlinear optical process, Opt. Lett. 2010. Vol. 35. P. 1581–1583.

30. Khonina S.N. Specular and vortical Airy beams, Opt. Commun. 2011. Vol. 284. P. 4263–4271.

31. Hu Y., Zhang P., Lou C., Huang S., Xu J., Chen Z. Optimal control of the ballistic motion of Airy beams, Opt. Lett. 2010. Vol. 35. P. 2260–2263.

32. Carvalho M.I., Facao M. Propagation of Airy-related beams, Opt. Express. 2010. Vol. 18. P. 21938–21949.

33. Efremidis N.K., Christodoulides D.N. Abruptly autofocusing waves, Opt. Lett. 2010. Vol. 35. P. 4045–4047.

34. Longhi S. Airy beams from a microchip laser, Opt. Lett. 2011. Vol. 36. P. 716–718.

35. Abdollahpour D., Suntsov S., Parazoglou D.G., Tzortzakis S. Spatiotemporal Airy light bullets in the linear and nonlinear regimes, Phys. Rev. Lett. 2010. Vol. 105. P. 253901.

36. Greenfield E., Segev M., Walasik W., Raz O. Accelerating light beams along arbitrary convex trajectories, Phys. Rev. Lett. 2011. Vol. 106. P. 213902.

37. Efremidis N.K. Airy trajectory engineering in dynamic linear index potential, Opt. Lett. 2011. Vol. 36. P. 3006–3008.

38. Froehly L., Courvoisier F., Mathis A., Jacquot M., Furfaro L., Giust R., Lacourt P.A., Dudley J.M. Arbitrary accelerating micron-scale caustic beams in two and three dimensions, Opt. Express. 2011. Vol. 19. P. 16455–16465.

39. Ye Z., Liu S., Lou C., Zhang P., Hu Y., Song D., Zhao J., Chen Z. Acceleration control of Airy beams with optically induced refractive-index gradient, Opt. Lett. 2011. Vol. 36. P. 3230–3232.

40. Li L., Li T., Wabg S.M., Zhang C., Zhu S.N. Plasmonic Airy beam generated by in-plane diffraction, Phys. Rev. Lett. 2011. Vol. 107. P. 126804.

41. Cottrell D.M., Devis J.A., Hazard T.M. Direct generation of accelerating Airy beams using a 3/2 phase-only pattern, Opt. Lett. 2009. Vol. 34. P. 2634–2636.

42. Abramochkin E.G., Razueva E. Product of three Airy beams, Opt. Lett. 2011. Vol. 36. P. 3732–3734.

43. Porat G., Dolev I., Barlev O., Arie A. Airy beam laser, Opt. Lett. 2011. Vol. 36. P. 4119–4121.

44. Kaminer I., Bekenstein R., Nemirovsky J., Segev M. Nondiffracting accelerating wave packets of Maxwell's equations, Phys. Rev. Lett. 2012. Vol. 108. P. 163901.

45. Zhang P., Hu Y., Cannan D., Salandrino A., Li T., Morandotti R., Zhang X., Chen Z. Generation of linear and nonlinear nonparaxial accelerating beams, Opt. Lett. 2012. Vol. 37. P. 2820–2822.

46. Courvoisier F., Mathis A., Froehly L., Giust R., Furfaro R., Lacourt P.A., Jacquot M., Dudley J.M. Sending femtosecond pulses in circles: highly nonparaxial accelerating beams, Opt. Lett. 2012. Vol. 37. P. 1736–1738.

47. Jiang Y., Huang K., Lu X. Propagation dynamics of abruptly autofocusing Airy beams with optical vortices, Opt. Express. 2012. Vol. 20. P. 18579–18584.

48. Yan S., Yao B., Lei M., Dan D., Yang Y., Gao P. Virtual source for an Airy beam, Opt. Lett. 2012. Vol. 37. P. 4774–4776.

49. Aleahmad P., Miri M., Mills M.S., Kaminer I. Fully vectoral accelerating diffraction-free Helmholtz beams, Phys. Rev. Lett. 2012. Vol. 109. P. 203902.

50. Kotlyar V.V., Kovalev A.A. Airy beam with a hyperbolic trajectory, Opt. Commun. 2014. Vol. 313. P. 290–293.

51. Khonina S.N., Volotovsky S.G. Bounded one-dimensional Airy beams: laser fan. Computer Optics. 2008. V.32. No.2. P.168-175.

52. Zamboni-Rached M., Nobrega K.N., Dartora C.A. Analytic description of Airy-type beams when truncated by finite apertures, Opt. Express. 2012. Vol. 20, No. 18. P. 19972–19977.

53. Abramovitz M., Stegun I.A. Handbook of mathematical functions. - Dover Publications, 1965.

54. Kaganovsky Y., Heyman E. Wave analysis of Airy beams, Opt. Express. 2010. Vol. 18, No. 8. P. 8440–8452.

55. Bandres M.A., Rodriguez-Lara B.M. Nondiffracting accelerating waves: Weber waves and parabolic momentum, New J. Phys. 2013. Vol. 15. P. 013054.

56. Torre A. A note on the general solution of the paraxial wave equation: a Lie algebra view, J. Opt. A: Pure Appl. Opt. 2008. Vol. 10. P. 055006.

57. Bandres M.A., Alonso M.A., Kaminer I., Segev M. Three-dimensional accelerating electromagnetic waves, Opt. Express. 2013. Vol. 21. No. 12. P. 13917–13929.

58. Kotlyar V.V., Stafeev S.S., Kovalev A.A. Curved laser microjet in near field, Appl. Opt. 2013. Vol. 52, No. 18. P. 4131–4136.

59. Born M., Wolf E., Principles of optics. Pergamon Press, 1973.

60. Kogelnik H., Li T. Laser beams and resonators, Proc. IEEE. 1966. Vol. 54. P. 1312–1329.

61. Kotlyar V.V., Kovalev A.A. Orbital angular momentum of superposition of two generalized Hermite–Gaussian laser beams. Computer Optics. 2013. V.37. No.2. P.179-185.

62. Kotlyar V.V., Kovalev A.A., Nalimov A.G. Propagation of hypergeometric laser beams in a medium with the parabolic refractive index, J. Opt. 2013. Vol. 313. P. 290–293.

63. Kotlyar V.V., Kovalev A.A. Family of hypergeometric laser beams, J. Opt. Soc. Am. A. 2008. Vol. 25, No. 1. P. 262–270.

64. Papazoglou D.G., Efremidis N.K., Christodoulides D.N., Tzortakis S. Observation of abruptly autofocusing waves, Opt. Lett. 2011. Vol. 32, No. 10. P. 1842–1844.

65. Prudnikov A.P., Brychkov A.P., Marichev O.I. Integrals and series. Special functions. Moscow, Nauka, 1983.

66. Kotlyar V.V., Kovalev A.A., Soifer V.A. Transformation of decelerating laser beams into accelerating ones, J. Opt. 2014. Vol. 16, No. 8. P. 085701.

67. Diffractive Nanophotonics. Ed. V.A. Soifer. Moscow. Fizmatlit. 2011.

68. Luneburg R.K. Mathematical Theory of Optics. Berkeley: University of California Press, 1966.

69. Zhang Y., Wang L., Zheng C. Vector propagation of radially polarized Gaussian beams diffracted by an axicon, J. Opt. Soc. Am. A. 2005. Vol. 22. P. 2542–2546.

70. Ozaktas H., Koc, A., Sari I., Kutay M. Efficient computation of quadratic-phase integrals in optics, Opt. Lett. 2006. Vol. 31. P. 35–37.

71. Koc, A., Ozaktas H., Hesselink L. Fast and accurate computation of two-dimensional non-separable quadratic-phase integrals, J. Opt. Soc. Am. A. 2010. Vol. 27. P. 1288–1302.

72. Kotlyar V.V., Kovalev A.A., Nalimov A.G. Hypergeometric laser beams in a parabolic waveguide. Computer Optics. 2012. V. 36. No. 3. P. 308–315.

73. Khonina S.N., Striletz A.S., Kovalev A.A., Kotlyar V.V. Propagation of laser vortex beams in a parabolic optical fiber, Proc. SPIE. 2009. Vol. 7523. P. 7523B.

74. Striletz A.S., Khonina S.N. Review and study of methods based on differential and integral operators of laser propagation in media with small inhomogeneities. Computer Optics. 2008. V. 32. No. 1. P.33–38.

75. Touam T., Yergeau F. Analytical solution for a linearly graded-index-profile planar waveguide, Appl. Opt. 1993. Vol. 32. P. 309–312.

76. Neganov V.A., Raevskii S.B., Yarovoi G.P. Linear macroscopic electrodynamics, Moscow, Radio i svyaz'. 2000.

77. Vaveliuk P., Lencina A., Rodrigo J., Matos O. Symmetric Airy beams, Opt. Lett. 2014. Vol. 39, No. 8. P. 2370–2373.

78. Chu X., Wen W. Effect of aberrations on the parameters of an Airy beam, Opt. Commun. 2014. Vol. 324. P. 18–21.

79. Liang Y., Hu Y., Ye Z., Song D., Lou C., Zhang X., Xu J., Morandotti R., Chen Z. Dynamical deformed Airy beams with arbitrary angles between two wings, J. Opt. Soc. Am. A. 2014. Vol. 31, No. 7. P. 1468–1472.

80. Li P., Liu S., Peng T., Xie G., Gan X., Zhao J. Spiral autofocusing Airy beams carrying power-exponent-phase vortices, Opt. Express. 2014. Vol. 22, No. 7. P. 7598–7606.

81. Ruelas A., Davis J., Moreno I., Cottrell D., Bandres M. Accelerating light beams with arbitrarily transverse shapes, Opt. Express 2014. Vol. 22, No. 3. P. 3490–3500.

82. Zhang P., Hu Y., Li T., Cannan D., Yin X., Morandotti R., Chen Z., Zhang X. Nonparaxial Mathieu and Weber Accelerating Beams, Phys. Rev. Lett. 2012. Vol. 109. P. 193901.

83. Alonso M., Bandres M. Generation of nonparaxial accelerating fields through mirrors. I: Two dimensions, Opt. Express. 2014. Vol. 22, No. 6. P. 7124–7132.

84. Kotlyar V., Kovalev A., Soifer V., Tuvey C., Davis J. Sidelobe contrast reduction for optical vortex beams using a helical axicon, Opt. Lett. 2007. Vol. 32. P. 921–923.

85. Ohtake Y., Ando T., Inoue T., Matsumoto N., Toyoda H. Sidelobe reduction of tightly focused radially higher-order Laguerre–Gaussian beams using annular masks, Opt. Lett. 2008. Vol. 33. P. 617–619.

86. Barwick S. Reduced side-lobe Airy beams, Opt. Lett. 2011. Vol. 36. P. 2827–2829.

87. Kotlyar V.V., Kovalev A.A., Soifer V.A. Diffraction-free asymmetric elegant Bessel beams with fractional orbital angular momentum. Computer Optics. 2014. V. 38. No.1. P.4-10.

88. Kotlyar V.V., Kovalev A.A., Soifer V.A. Asymmetric Bessel modes, Opt. Lett. 2014.

Vol. 39, No. 8. P. 2395–2398.

89.   Lombardo K., Alonso M.A. Orthonormal basis for nonparaxial focused fields in two dimensions, and its application to modeling scattering and optical manipulation of objects, Am. J. Phys. 2012. Vol. 80, No. 1. P. 82–93.

90.   Kravtsov Yu.A. Complex ray and complex caustics, Radiophys. Quantum. Electron. 1967. Vol. 10. P. 719–730.

91.   Arnaud J. Degenerate optical cavities. II: Effects of misalignments, Appl. Opt. 1967. Vol. 8. P. 1909–1917.

92.   Keller J.B., Streifer W. Complex rays with an application to Gaussian beams, J. Opt. Soc. Am. 1971. Vol. 61. P. 40–43.

93.   Deschamps G.A. Gaussian beams as a bundle of complex rays, Electron. Lett. 1971. Vol. 7. P. 684–685.

94.   Berry M., Nye J., Wright F. The elliptic umbilic diffraction catastrophe, Phil. Trans. R. Soc. Lond. 1979. Vol. 291. P. 453–484.

95.   Vaughan J.M., Willetts D. Interference properties of a light-beam having a helical wave surface, Opt. Commun. 1979. Vol. 30. P. 263–267.

96.   Coullet P., Gil G., Rocca F. Optical vortices, Opt. Commun. 1989. Vol. 73. P. 403–408.

97.   Bazhenov V., Vasnetsov M.V., Soskin M.S. Laser-beam with screw dislocations in the wavefront, JETP Lett. 1990. Vol. 52. P. 429–431.

98.   Khonina S.N., Kotlyar V.V., Shinkarev M.V., Soifer V.A., Uspleniev G.V. The rotor phase filter, J. Mod. Opt. 1992. Vol. 39. P. 1147–1154.

99.   Abramochkin E.G., Volostnikov V.G. Beam transformation and nontransformed beams, Opt. Commun. 1991. Vol. 83. P. 123–125.

100. Beijersbergen M.W., Allen L., Van der Veen H.E., Woerdman J.P. Astigmatic laser mode converters and transfer of orbital angular momentum, Opt. Commun. 1993. Vol. 96. P. 123–132.

101. Yao A.M., Padgett M.J. Orbital angular momentum: origins, behavior and applications, Adv. Opt. Photon. 2011. Vol. 3. P. 161–204.

102. Vyas S., Senthilkumaran P. Interferometric optical vortex array generator, Appl. Opt. 2007. Vol. 46. P. 2893–2898.

103. Fraczek E., Budzyn G. An analysis of an optical vortices interferometer with focused beam, Opt. Applicata. 2009. Vol. XXXIX. P. 91–99.

104. Singh B.K., Singh G., Senthilkumaran P., Metha D.S. Generation of optical vortex array using single-element reversed-wavefront folding interferometer. Int. J. Opt. 2012. Vol. 2012. P. 689612.

105. Shen Y., Campbell G.T., Hage B., Zou H., Buchler B.C., Lam P.K. Generation and interferometric analysis of high charge optical vortices, J. Opt. 2013. Vol. 15. P. 044005.

106. Gotte J.B., O'Holleran K., Precce D., Flossman F., Franke-Arnold S., Barnett S.M., Padgett M.J. Light beams with fractional orbital angular momentum and their vortex structure, Opt. Express. 2008. Vol. 16. P. 993–1006.

107. O'Dwyer D.P., Phelan C.F., Rakovich Y.P., Eastham P.R., Lunney J.C., Donegan J.F. Generation of continuously tunable fractional orbital angular momentum using internal conical diffraction, Opt. Express. 2010. Vol. 18. P. 16480–16485.

108. Siegman A.E. Hermite-Gaussian functions of complex argument as optical beam eigenfunction, J. Opt. Soc. Am. 1973. Vol. 63. P. 1093–1094.

109. Khonina S.N., Kotlyar V.V., Soifer V.A., Paakkonen P., Simonen J., Turunen J. An analysis of the angular momentum of a light field in terms of angular harmonics, J. Mod. Opt. 2001. Vol. 48. P. 1543–1557.

110. Kotlyar V.V., Kovalev A.A. Hermite-Gaussian modal laser beams with orbital angular momentum, J. Opt. Soc. Am. A. 2014. Vol. 31, No. 2. P. 274–282.

111. Pratesi R., Ronchi L. Generalized Gaussian beams in free space, J. Opt. Soc. Am. A. 1977. Vol. 67. P. 1274–1276.

112. Gradshteyn I.S., Ryzhik I.M. Table of Integrals, Series, and Products: 5th edition. New York: Academic, 1996. 1762 p.

113. Magnus W., Oberhettinger F., Sony R.P. Formulas and theorems for the special functions of mathematical physics: Third Edition. Springer-Verlag Berlin Heidelberg, 1966. 508 p.

114. Miller Jr. W. Symetry and separation of variables. Addison-Wesley Pub. Comp., 1977.

115. Kotlyar V.V., Khonina S.N., Soifer V.A. Algorithm for the generation of non-diffracting Bessel beams, J. Mod. Opt. 1992. Vol. 42, No. 6. P. 1231–1239.

116. Gutiérrez-Vega J.C., Iturbe-Castillo M.D., Chavez-Cedra S. Alternative formulation for invariant optical fields: Mathieu beams, Opt. Lett. 2000. Vol. 25, No. 20. P. 1493–1495.

117. Chávez-Cedra S., Gutiérrez-Vega J.C., New G.H.C. Elliptic vortices of electromagnetic wave fields, Opt. Lett. 2001. Vol. 26, No. 22. P. 1803–1805.

118. Dennis M.R., Ring J.D. Propagation-invariant beams with quantum pendulum spectra: from Bessel beams to Gaussian beams, Opt. Lett. 2013. Vol. 38, No. 17. P. 3325–3328.

119. Gori F., Guattari G., Padovani C. Bessel-Gauss beams, Opt. Commun. 1987. Vol. 64. P. 491–495.

120. Li Y., Lee H., Wolf E. New generalized Bessel-Gauss beams, J. Opt. Soc. Am. A. 2004. Vol. 21. P. 640–646.

121. Kisilev A.P. New structures in paraxial Gaussian beams, Opt. Spectrosc. 2004. Vol. 96. P. 479–481.

122. Gutiérrez-Vega J.C., Bandres M.A. Helmholz-Gauss waves, J. Opt. Soc. Am. A. 2005. Vol. 22. P. 289–298.

123. Kotlyar V.V., Kovalev Skidanov R.V., A.A., Soifer V.A. Rotating elegant Bessel-Gaussian beams. Computer Optics. 2014. V. 38. No. 2. P. 162–170.

124. Kotlyar V.V., Kovalev A.A., Skidanov R.V., Soifer V.A. Asymmetric Bessel-Gauss beams, J. Opt. Soc. Am. A. 2014. Vol. 31, No. 9. P. 1977–1983.

125. Durnin J. Exact solution for nondiffractive beams. I. The scalar theory, J. Opt. Soc. Am. A. 1987. Vol. 4, No. 4. P. 651–654.

126. Turunen J., Vasara A., Friberg A.T. Holographic generation of diffractive-free beams, Appl. Opt. 1988. Vol. 27, No. 19. P. 3959–3962.

127. Vasara A., Turunen J., Friberg A.T. Realization of general nondiffracting beams with computer-generated holograms, J. Opt. Soc. Am. A. 1989. Vol. 6, No. 11. P. 1748–1754.

128. MacDonald R.P., Boothroyd S.A., Okamato T., Chrostowski J., Syrett B.A. Interboard optical data distribution by Bessel beam shadowing, Opt. Commun. 1996. Vol. 122. P. 169–177.

129. McQueen C.A., Arlt J., Dholakia K. An experiment to study a nondiffracting light beam, Am. J. Phys. 1999. Vol. 67. P. 912–915.

130. Barnett S.M., Allen L. Orbital angular momentum and nonparaxial light-beams, Opt. Commun. 1994. Vol. 110. P. 670–678.

131. Volke-Sepulveda K., Garcés-Chávez V., Chávez-Cedra S. et al. Orbital angular momentum of a high-order Bessel light beam, J. Opt. B-Quantum S. O. 2002. Vol. 4. P. S82–S89.

132. Kotlyar V.V., Khonina S.N., Soifer V.A. An algorithm for the generation of laser beams with londitudinal periodicity: rotating images, J. Mod. Opt. 1997. Vol. 44, No. 7. P. 1409–1416.

133. Paakkonen P., Lautanen J., Honkanen M. et al. Rotating optical fields: experimental demonstration with diffractive optics, J. Mod. Opt. 1998. Vol. 45, No. 11. P. 2355–2369.

134. Khonina S.N., Kotlyar V.V., Soifer V.A., Lautanen J., Honkanen M., Turunen J. Generating a couple of rotating nondiffracting beams using a binary-phase DOE, Optik. 1999. Vol. 110, No. 3. P. 137–144.

135. Lee H.S., Stewart B.W., Choi K., Fenichel H. Holographic nondiverging hollow beam, Phys. Rev. A. 1994. Vol. 49, No. 6. P. 4922–4927.

136. Herman R.M., Wiggins T.A. Production and uses of diffractionless beams, J. Opt. Soc. Am. A. 1991. Vol. 8, No. 6. P. 932–942.

137. Arlt J., Dholakia K. Generation of high-order Bessel beams by use of an axicon, Opt. Commun. 2000. Vol. 177. P. 297–301.

138. Kotlyar V.V., Khonina S.N., Soifer V.A., Uspleniev G.V., Shinkarev M.V. Trochoson, Opt. Commun. 1992. Vol. 91, No. 3-4. P. 158–162.

139. Devis J.A., Carcole E., Cottrell D.M. Intensity and phase measurements of nondiffracting beams generated with a magneto-optic spatial light modulator. Appl. Opt. 1996. Vol. 35, No. 4. P. 593–598.

140. Khonina S.N., Kotlyar V.V., Soifer V.A., Jefimovs K., Paakkonen P., Turunen J. Astigmatic Bessel laser beams, J. Mod. Opt. 2004. Vol. 51, No. 5. P. 677–686.

141. MacDonald M.P., Paterson L., Volke-Sepulveda K., Arlt J., Sibbett W., Dholakia K. Creation and manipulation of three-dimensional optically trapped structures, Science. 2002. Vol. 296. P. 1101–1103.

142. Garces-Chávez V., McGloin D., Melville H., Sibbett W., Dholakia K. Simultaneous micromanipulation in multiple planes using a self-reconstructing light beam, Nature. 2002. Vol. 419. P. 145–147.

143. Khonina S.N., Kotlyar V.V., Skidanov R.V., Soifer V.A., Jefimovs K., Simonen J., Turunen J. Rotation of microparticles with Bessel beams generated by diffractive elements, J. Mod. Opt. 2004. Vol. 51, No. 14. P. 2167–2184.

144. Arlt J., Hitomi T., Dholakia K. Atom guiding long Laguerre-Gaussian and Bessel beams, Appl. Phys. B-Lasers O. 2000. Vol. 71, Iss. 4. P. 549–554.

145. Arlt J., Dholakia K., Soneson J., Wright E.M. Optical dipole traps and atomic waveguides based on Bessel light beams, Phys. Rev. A. 2001. Vol. 63. P. 063602.

146. Kotlyar V.V., Kovalev A.A., Soifer V.A. Hankel-Bessel laser beams, J. Opt. Soc. Am. A. 2012. Vol. 29, No. 5. P. 741–747.

147. Zhu Y., Liu X., Gao J., Zhang Y., Zhao F. Probability density of the orbital angular momentum mode of Hankel-Bessel beams in an atmospheric turbulence, Opt. Express. 2014. Vol. 22, No. 7. P. 7765–7772.

148. Bouchal Z., Olivik M. Non-diffractive vector Bessel beams, J. Mod. Opt. 1995. Vol. 42, No. 8. P. 1555–1566.

149. Yu Y.Z., Dou W.B. Vector analysis of nondiffracting Bessel beams, Prog. in Electr. Res. Lett. 2008. Vol. 5. P. 57–71.

150. Litvin I.A., Dudley A., Forbes A. Poynting vector and orbital angular momentum density of superpositions of Bessel beams, Opt. Express. 2011. Vol. 19, No. 18. P. 16760–16771.

151. Chen Y.F., Lin Y.C., Zhuang W.Z., Liang H.C., Su K.W., Huang K.F. Generation of large orbital angular momentum from superposed Bessel beams corresponding to resonant geometric modes, Phys. Rev. A. 2012. Vol. 85. P. 043833.

152. Gong L., Qui X., Ren Y., Zhu H., Liu W., Zhou J. Observation of the asymmetric Bessel bems with arbitrary orientation using a digital micromirror device, Opt. Express. 2014. Vol. 22, No. 22. P. 26763–26776.

153. Sheppard C.J.R., Kou S.S., Lin J. Two-dimensional complex source point solutions: application to propagationally invariant beams, optical fiber modes, planar waveguides, and plasmonic devices, J. Opt. Soc. Am. A. 2014. Vol. 31, No. 12. P. 2674–2679.

154. Kovalev A.A., Kotlyar V.V., Porfirev A.P. Shifted nondiffractive Bessel beams, Phys. Rev. A. 2015. Vol. 91. P. 053840.

155. Kovalev A.A., Kotlyar V.V., Porfirev A.P., Kalinkina D.S. Research of orbital angular momentum of superpositions of diffraction-free Bessel beams with a complex shift. Computer Optics. 2015. V. 39. No. 2. P. 172–180.

156. Mitri F.G. Partial-wave series expansion in spherical coordinates for the acoustic field of vortex beams generated from finite circular aperture, IEEE T. Ultrason. Ferr. 2014. Vol. 61, No. 12. P. 2089–2097.

157. Mendez G., Fernando-Vazquez A., Lopez R.P. Orbital angular momentum and highly efficient holographic generation of nondiffractive TE and TM vector beams, Opt. Commun. 2015. Vol. 334. P. 174–183.

158. Kotlyar V.V., Kovalev A.A., Soifer V.A. Superpositions of asymmetrical Bessel beams, J. Opt. Soc. Am. A. 2015. Vol. 32, No. 6. P. 1046–1052.

159. Courtial J., Zambrini R., Dennis M., Vasnetsov M. Angular momentum of optical vortex arrays, Opt. Express. 2006. Vol. 4. P. 938–949.

160. Martinez-Castellanos I., Gutierrez-Vega J. Shaping optical beams with non-integer orbital angular momentum: a generalized differential operator approach, Opt. Lett. 2015. Vol. 40, No. 8. P. 1764–1767.

161. Kovalev A.A., Kotlyar V.V. Orbital angular momentum of superposition of identical shifted vortex beams, J. Opt. Soc. Am. A. 2015. Vol. 32, No. 10. P.1805–1810.

162. Lu J., Greenleaf J. Diffraction-limited beams and their applications for ultrasonic imaging and tissue characterization, Am. J. Phys. 1979. Vol. 47, No. 3. P. 264–267.

163. Nelson W., Palastro J.P., Davis C.C., Sprangle P. Propagation of Bessel and Airy beams through atmospheric turbulence, J. Opt. Soc. Am. A. 2014. Vol. 31, No. 3. P. 603–609.

164. Froehly L., Jacquot M., Lacourt P.A., Dudley J.M., Courvoisier F. Spatiotemporal structure of femtosecond bessel beams from spatial light modulators. J. Opt. Soc. Am. A. 2014. Vol. 31, No. 4. P. 790–793.

165. Watson G.N. A Treatise on the Theory of Bessel Functions: 2nd ed. Cambridge, England: Cambridge University Press, 1966. . 16.5-16.59. P. 537–550.

166. Sheppard C.J.R. Focusing of vortex beams: Lommel treatment, J. Opt. Soc. Am. A. 2014. Vol. 31, No. 3. P. 644–651.

167. Skidanov R.V., Ganchevskaya S.V. Diffractive optical elements for the formation of combinations of vortex beams in the problem manipulation of microobjects. Computer Optics. 2014. V. 38. No. 1. P. 65–71.

168. Kovalev A.A., Kotlyar V.V. Lommel modes, Opt. Commun. 2015. Vol. 338. P. 117–122.

169. Huang S., Miao Z., He C., Pang F., Li Y., Wang T. Composite vortex beams by co-axial superposition of Laguerre–Gaussian beams, Opt. Las. Eng. 2016. Vol. 78. P. 132–139.

170. Plick W.N., Krenn M. Physical meaning of the radial index of Laguerre–Gauss beams, Phys. Rev. A. 2015. Vol. 92, No. 6. P. 063841.

171. Savelyev D.A., Khonina S.N. Characteristics of sharp focusing of vortex Laguerre-

Gaussian beams, Computer Optics. 2015. Vol. 39, No. 5. P. 654–662.

172. Stilgoe A.B., Nieminen T.A., Rubinsztein-Dunlop H. Energy, momentum and prop-agation of non-paraxial high-order Gaussian beams in the presence of an aperture, J. Opt. 2015. Vol. 17, No. 12. P. 125601.

173. Zhang Y., Liu X., Belizh M., Zhong W., Wen F., Zhang Y. Anharmonic propagation of two-dimensional beams carrying orbital angular momentum in a harmonic poten-tial, Opt. Lett. 2015. Vol. 40. P. 3786–3789.

174. Kim D.J., Kim J.W. High-power TEM00 and Laguerre–Gaussian mode generation in double resonator configuration, Appl. Phys. B. 2015. Vol. 121, No. 3. P. 401–405.

175. Lin D., Daniel J., Clarkson W. Controlling the handedness of directly excited Laguerre-Gaussian modes in a solid-state laser, Opt. Lett. 2014. Vol. 39. P. 3903–3906.

176. Ruffato G., Massari M., Romanato F. Generation of high-order Laguerre-Gaussian modes by means of spiral phase plates, Opt. Lett. 2014. Vol. 39. P. 5094–5097.

177. Das B.C., Bhattacharyya D., De S. Narrowing of Doppler and hyperfine line shapes of Rb–D2 transition using a vortex beam, Chem. Phys. Lett. 2016. Vol. 644. P. 212–218.

178. Allocca A., Gatto A., Tacca M., Day R.A., Barsuglia M., Pillant G., Buy C., Vajente G. Higher-order Laguerre-Gauss interferometry for gravitational-wave detectors with in situ mirror defects compensation, Phys. Rev. D. 2015. Vol. 92, No. 10. P. 102002.

179. Sun K., Qu C., Zhang C. Spin-orbital-angular-momentum coupling in Bose-Einstein condensates, Phys. Rev. A. 2015. Vol. 91, No. 6. P. 063627.

180. Mondal P.K., Deb B., Majumder S. Angular momentum transfer in interaction of Laguerre-Gaussian beams with atoms and molecules, Phys. Rev. A. 2014. Vol. 89, No. 6. P. 063418.

181. Otsu T., Ando T., Takiguchi Y., Ohtake Y., Toyoda H., Itoh H. Direct evidence for three-dimensional off-axis trapping with single Laguerre-Gaussian beam, Sci. Rep. 2014. Vol. 4. P. 4579.

182. Krenn M., Fickler R., Fink M., Handsteiner J., Malik M., Scheidl T., Ursin R., Zeil-inger A. Communication with spatially modulated light through turbulent air across Vienna, New J. Phys. 2014. Vol. 16. P. 113028.

183. Kravtsov Yu.A. Complex ray and complex caustics, Radiophys. Quantum Electron. 1967. Vol. 10. P. 719–730.

184. Kim H.C., Lee Y.H. Hermite–Gaussian and Laguerre–Gaussian beams beyond the paraxial approximation, Opt. Commun. 1999. Vol. 169. P. 9–16.

185. Rykov M.A., Skidanov R.V. Modifying the laser beam intensity distribution for ob-taining improved strength characteristics of an optical trap, Appl. Opt. 2014. Vol. 53, No. 2. P. 156–164.

186. Mair A., Vaziri A., Weihs G., Zeilinger A. Entanglement of the orbital angular mo-mentum states of photons, Nature. 2001. Vol. 412, No. 6844. P. 313–316.

187. Kovalev A.A., Kotlyar V.V., Porfirev A.P. Asymmetric Laguerre-Gaussian beams, Phys. Rev. A. 2016. Vol. 93. P. 063858.

188. Kovalev A.A., Kotlyar V.V., Porfirev A.P. Optical trapping and moving of micropar-ticles by using asymmetric Laguerre-Gaussian beams, Opt. Lett. 2016. Vol. 41, No. 11. P.2426–2429.

189. Kotlyar V.V., Almazov A.A., Khonina S.N., Soifer V.A., Elfstrom H., Turunen J. Generation of phase singularity through diffracting a plane or Gaussian beam by a spiral phase plate, J. Opt. Soc. Am. A. 2005. Vol. 22, No. 5. P. 849–861.

190. Pearcey T. The structure of an electromagnetic field in the neighbourhood of a cusp

of a caustic, Phil. Mag. S. 1946. Vol. 7, No. 37. P. 311–317.

191. Berry M.V., Howls C.J. Integrals with coalescing saddles, http://dlmf.nist.gov/36.2 (Digital Library of Mathematical Functions, National Institute of Standards and Technology, 2012).

192. Deng D., Chen C., Zhao X., Chen B., Peng X., Zheng Y. Virtual Source of a Pearcey beam, Opt. Lett. 2014. Vol. 39, No. 9. P. 2703–2706.

193. Ronzitti E., Guillon M., de Sars V., Emiliani V. L. CoS nematic SLM characterization and modeling for diffraction efficiency optimization, zero and ghost orders suppression, Opt. Express. 2012. Vol. 20, No. 16. P. 17843–17855.

194. Kotlyar V.V., Kovalev A.A., Porfirev A.P. Generation of half Pearcey laser beams by a spatial light modulator. Computer Optics. V. 38. No. 4. P. 658–662.

195. Kovalev A.A., Kotlyar V.V., Zaskanov S.G., Profirev A.P. Half Pearcey laser beams, J. Opt. 2015. Vol. 17. P. 035604.

196. Wang J., Yang J.-Y., Fazal I.M. et al. Terabit free-space data transmission employing orbital angular momentum multiplexing, Nat. Photon. 2012. Vol. 6. P. 488–496.

197. Khonina S.N., Kotlyar V.V., Skidanov R.V., Soifer V.A., Jefimov K., Simonen J., Turunen J. Rotation of microparticles with Bessel beams generated by diffractive elements, J. Mod. Opt. 2004. Vol. 51, No. 14. P. 2167–2184.

198. Zhu Y., Liu X., Cao J., Zhang Y., Zhao F. Probability density of the orbital angular momentum mode of Hankel-Bessel beams in an atmospheric turbulence, Opt. Express. 2014. Vol. 22, No. 7. P. 7765–7772.

199. Kovalev A.A., Kotlyar V.V., Zaskanov S.G. Structurally stable three-dimensional and two-dimensional laser half Pearcey beams. Computer Optics. 2014. V. 38. No. 2. P. 193–197.

200. Kovalev A.A., Kotlyar V.V. Pearcey beams carrying orbital angular momentum. Computer Optics. 2015. V. 39. No. 4. P. 453–458.

201. Kovalev A.A., Kotlyar V.V., Porfirev A.P. Auto-focusing accelerating hyper-geometric laser beams. Journal of Optics, v.18, no.2, p. 025610 (2016).

202. Karimi E., Zito G., Piccirillo B., Marrucci L., Santameto E. Hypergeometric Gaussian modes, Opt. Lett. 2007. Vol. 32. P. 3053–3055.

203. Kotlyar V.V., Kovalev A.A., Skidanov R.V., Khonina S.N., Turunen J. Generating hypergeometric laser beams with a diffractive optical elements, Appl. Opt. 2008. Vol. 47, No. 32. P. 6124–6133.

204. Chen J., Wang G., Xu Q. Production of confluent hypergeometric beam by computer-generated hologram, Opt. Eng. 2011. Vol. 50, No. 2. P. 024201.

205. Bernardo B., Moraes F. Data transmission by hypergeometric modes through a hyperbolic-index medium, Opt. Exp. 2011. Vol. 19, No. 2. P. 11264–11270.

206. Li J., Chen Y. Propagation of confluent hypergeometric beam through unaxial crystals orthogonal to the optical axis, Opt. Las. Technol. 2012. Vol. 44. P. 1603–1610.

207. Di Falco A., Kehr S.C., Leonhardt U. Luneberg lens in silicon photonics, Opt. Express 2011. Vol. 19. P. 5156–5162.

208. Zentgrat T., Liu Y., Mikkelsen M.N., Valentine J., Zhang X. Plasmonic Luneberg and Eaton lenses, Nat. Nanotechn. 2011. Vol. 6. P. 151–155.

209. Dyachenko P.N., Pavelyev V.S., Soifer V.A. Graded photonic quasicrystals. Opt. Lett. 2012. Vol. 37, No. 12. P. 2178–2180.

210. Triandaphilov Y.R., Kotlyar V.V. Photonic Crystal Mikaelian Lens, Opt. Mem. Neur. Netw. (Inform. Opt.). 2008. Vol. 17. P. 1–7.

211. Kotlyar V.V., Kovalev A.A., Soifer V.A. Subwavelength Focusing with a Mikaelian Planar Lens, Opt. Mem. Neur. Netw. (Inform. Opt.). 2010. Vol. 19. P. 273–278.

212. Kotlyar V.V., Stafeev S.S. Sharply focusing a radially polarized laser beam using a

gradient Mikaelian's microlens, Opt. Commun. 2009. Vol. 282, No. 4. P.459–464.

213.   Kotlyar V.V., Kovalev A.A., Nalimov A.G., Triandafilov Y.R. Mechanism of super-resolution in a planar hyperbolic secant lens. Computer Optics. 2010. V. 34. No. 4. P. 428–435.

214.   Kotlyar V.V., Khonina S.N., Wang Ya. Operator description of paraxial light fields. Computer optics. 2001. V. 21. P. 45–52.

215.   Mendlovic D., Ozaktas H.M. Fractional Fourier transform and their optical implementation: I., J. Opt. Soc. Am. A. 1993. Vol. 10, No. 9. P. 1875–1881.

216.   Lohmann A.W. Image rotation, Wigner rotation, and the fractional Fourier transform, J. Opt. Soc. Am. A. 1993. Vol. 10, No. 10. P. 2118–2186.

217.   Striletz A.S., Khonina S.N. Investigation of propagation of laser beams in a parabolic optical fiber with an integral paraxial operator. Computer Optics. 2007. V. 31. No. 4. P. 33–39.

218.   Dupuis A. Guiding in the visible with colorful solid-core Bragg fiber, Opt. Lett. 2007. Vol. 32. P. 2882–2884.

219.   Ostrovsky A.S., Rickenstorff-Parrao C., Arrizon V. Generation of the perfect optical vortex using a liquid-crystal spatial light modulator, Opt. Lett. 2013. Vol. 38, No. 4. P. 534–536.

220.   Chen M., Mazilu M., Arita Y., Wright E.M., Dholakia K. Dynamics of microparticles trapped in a perfect vortex beam, Opt. Lett. 2013. Vol. 38, No. 22. P. 4919–4922.

221.   Vaity P., Rusch L. Perfect vortex beam: Fourier transformation of a Bessel beam, Opt. Lett. 2015. Vol. 40. P. 597–600.

222.   Jabir M.V., Apurv Chaitanya N., Aadhi A., Samanta G.K. Generation of perfect vortex of variable size and its effect in angular spectrum of the down-converted photons, Sci. Rep. 2016. Vol. 6. P. 21877.

223.   García-García J., Rickenstorff-Parrao C., Ramos-García R., Arrizon V., Ostrovsky A. Simple technique for generating the perfect optical vortex, Opt. Lett. 2014. Vol. 39, No. 18. P. 5305–5308.

224.   Fedotowsky A., Lehovec K. Optimal filter design for annular imaging, Appl. Opt. 1974. Vol. 13, No. 12. P. 2919–2923.

225.   Korn G., Korn T. A handbook of mathematics for scientists and engineers. Moscow, Nauka, 1968

226.   Kotlyar V.V., Khonina S.N., Kovalev A.A., Soifer V.A., Elfstrom H., Turunen J. Diffraction of a plane, finite-radius wave by a spiral phase plate, Opt. Lett. 2006. Vol. 31, No. 11. P. 1597–1599.

227.   Kotlyar V.V., Kovalev A.A., Skidanov R.V., Moiseev O.Y., Soifer V.A. Diffraction of a finite-radius plane wave and a Gaussian beam by a helical axicon and a spiral phase plate, J. Opt. Soc. Am. A. 2007. Vol. 24, No. 7. P. 1955–1964.

228.   Kotlyar V.V., Kovalev A.A., Porfirev A.P. Generating a perfect optical vortex: comparison of approaches. Computer Optics. 2016. V. 40. No. 3. P. 312–321.

229.   Bandres M.A., Gutiérrez-Vega J.C., Chavez-Cedra S. Parabolic nondiffracting optical wave fields, Opt. Lett. 2004. Vol. 29, No. 1. P. 44–46.

230.   Cerjan A., Cerjan C. Orbital angular momentum of Laguerre–Gaussian beams beyond the paraxial approximation, J. Opt. Soc. Am. A. 2011. Vol. 28, No. 11. P. 2253–2260.

231.   Guo C.S., Liu X., He J.L., Wang H.T. Optimal annulus structures of optical vortices, Opt. Express. 2004. Vol. 12. P. 4625–4634.

232.   Guo J., Wei Z., Liu Y., Huang A. Analysis of optical vortices with suppressed sidelobes using modified Bessel-like function and traepzoid annulus modulation structures, J. Opt. Soc. Am. A. 2015. Vol. 32, No. 2. P. 195–203.

233. Kotlyar V.V., Kovalev A.A., Soifer V.A. Nonparaxial Hankel vortex beams of the first and second types. Computer Optics. 2015. V. 39. No. 3. P. 299–304.

234. Wang J., Yang J.-Y., Fazal I.M. et al. Terabit free-space data transmission employing orbital angular momentum multiplexing, Nat. Photon. 2012. Vol. 6. P. 488–496.

235. Khonina S.N., Kotlyar V.V., Skidanov R.V., Soifer V.A., Jefimov K., Simonen J., Turunen J. Rotation of microparticles with Bessel beams generated by diffractive elements, J. Mod. Opt. 2004. Vol. 51, No. 14. P. 2167–2184.

236. Zhu Y., Liu X., Cao J., Zhang Y., Zhao F. Probability density of the orbital angular momentum mode of Hankel–Bessel beams in an atmospheric turbulence, Opt. Express. 2014. Vol. 22, No. 7. P. 7765–7772.

237. Mishra S.R. A vector wave analysis of a Bessel beam, Opt. Commun. 1991. Vol.. 85. P. 159–161.

238. Bouchal Z., Olivık M. Non-diffractive vector Bessel beams, J. Mod. Opt. 1995. Vol. 42, No. 8. P. 1555–1566.

239. Horak R., Bouchal Z., Bajer J. Nondiffracting stationary electromagnetic fields, Opt. Commun. 1997. Vol. 133. P. 314–327.

240. Jauequi R., Hacyan S. Quantum-mechanical properties Bessel beams, Phys. Rev. A. 2005. Vol. 71. P. 033411.

241. Yu Y.Z., Dou W.B. Vector analyses of nondiffracting Bessel beams, Prog. in Electr. Res. Lett. 2008. Vol. 5. P. 57–71.

242. Wang Y., Dou W., Meng H. Vector analyses of linearly and circularly polarized Bessel beams using Herz vector potentials, Opt. Express. 2014. Vol. 22, No. 7. P. 7821–7830.

243. Mitri F.G. Vector spherical quasi-Gaussian vortex beams, Phys. Rev. E. 2014. Vol. 89. P. 023205.

244. Kotlyar V.V., Kovalev A.A. Vectorial Hankel laser beams carrying orbital angular momentum. Computer Optics. 2015. V. 39. No. 4. P. 449–452.

245. Kotlyar V.V., Kovalev A.A., Soifer V.A. Vectorial rotating vortex Hankel laser beams, J. Opt. 2016. Vol. 18, No. 9. P. 095602.

246. Youngworth K.S., Brown T.G. Focusing of high numerical aperture cylindrical-vector beams, Opt. Express. 2000. Vol. 7, No. 2. P. 77–87.

247. Zhan Q., Leger J.R. Focus shaping using cylindrical vector beams, Opt. Express. 2002. Vol. 10, No. 7. P. 324–331.

Index

V